MW00340124

100 INNOVATIONS OF
THE INDUSTRIAL
REVOLUTION

FROM 1700 TO 1860

© Simon Forty 2019

All rights reserved. No part of this publication may be reproduced or stored in a retrieval system or transmitted, in any form or by any means, electronic, mechanical, photocopying, recording or otherwise, without prior permission in writing from Haynes Publishing.

First published in April 2019

A catalogue record for this book is available from the British Library.

ISBN 978 1 78521 566 7

Library of Congress control no. 2018953081

Published by Haynes Publishing,
Sparkford, Yeovil,
Somerset BA22 7JJ, UK.
Tel: 01963 440635
Int. tel: +44 1963 440635
Website: www.haynes.com

Haynes North America Inc.
859 Lawrence Drive,
Newbury Park, California 91320, USA.

Printed in Malaysia.

Facing page: Boulton, Watt and Murdoch – three of Britain's greatest inventors – in Birmingham. Statue by William Bloye, 1956. Oosoom/WikiCommons (CC BY-SA 3.0)

ACKNOWLEDGEMENTS

This book is a team effort. It wouldn't have happened without the material help of three people: Peter Waller, Dr Patrick Hook and my wife Sandra. All three contributed materially, writing entries, helping with illustrations and providing information. Thanks, too, to Ian Kelly and Richard Wood who helped with the list.

The photographs came from a number of sources. I'd particularly like to thank Tom of https://www.whateversleft.co.uk for the photos of Stanley Mill. Thanks also to the Wellcome Collection (https://wellcomecollection.org/works) and the Library of Congress (http://www.loc.gov/pictures/), both of which provide fabulous online resources for researchers. Thanks, too, to the commercial libraries, in particular the Science Picture Library and Getty Images. The full list of photo credits can be found at the end of the book.

100 INNOVATIONS OF
THE INDUSTRIAL REVOLUTION

FROM 1700 TO 1860

SIMON FORTY

Contents

Preface

▼ *Coalbrookdale in the River Severn gorge in Shropshire was the thriving hub of the Industrial Revolution. The Iron Bridge is in the centre of this photo and the plume of white gases comes from the now-decommissioned power station further up the gorge at Buildwas.* English Heritage/Heritage Images/Getty Images

It's usually easy to pinpoint the duration of a lifespan: there's a birth and a death and a life in between. But influences and movements – and, surprisingly, inventions – are more difficult to tie down. Take as an example the Enlightenment, which overlaps and is indivisible from the Industrial Revolution. Does the so-called Age of Reason start in 1715 as some, notably the French, would have us believe, or earlier in the 17th century with the Scientific Revolution … or should that be 1543 when Copernicus published his *On the Revolutions of Heavenly Spheres*? Does Newton's *Principia* herald the end of the Scientific Revolution or the start of the Age of Reason? It's easy to go round in circles, and this is particularly true when we consider what has become known as the Industrial Revolution. Like Willi Brandt's famous dismissal of the DDR (*Die DDR sei*

'*weder deutsch, noch demokratisch, noch eine Republik*'), the Industrial Revolution often was neither industrial nor a revolution. It has no definable starting point and when did it end? As far as who invented what, the morass of patents, litigation and national pride ensures that even the seemingly obvious can mask a more subtle truth.

With all these questions it is the duty of authors to define their parameters. The starting point of this book was easy: first, the date – 1709; second, the person – Abraham Darby; third, the place – Coalbrookdale. What of the entries? There can be no correct or incorrect versions of an entry list that is so dependent on an author's whim, knowledge or interests. Too many railways? Not enough architecture? Too Anglophile? I apologise to those who may find omissions or think the content unbalanced. I have tried to combine mechanical advances (thanks to Dr Patrick Hook for his technical explanations) with elements visible today: the concrete examples of industrial archaeology that stud the British landscape.

As the first entry in the book explains, without iron most of the inventions and locations I've covered would not exist. Without coke – or in the first place coal – the iron that made many of the objects in this book – from bridges to bicycles – wouldn't have been strong or plentiful enough. Without steam power these ironworks would have had to use water power, which naturally fluctuated and rarely was consistently powerful enough. Five of the first eight entries in the book look at these subjects.

There are many other themes that run through the book and inform the choice of subjects: Britain in 1709 was an agrarian society. Animal and crop husbandry were the source of wealth for some and the livelihood of many. Unsurprisingly, improvements in this area play a large part in 18th-century innovations. Britain was also in thrall to the manufacture of textiles. This had been true for many years: the wool churches of the medieval Cotswolds and the great abbeys of Yorkshire were built on wool. The Lord Chancellor in the House of Lords sits on the Woolsack. Once the cotton industry had taken hold in Lancashire, it would dominate British productivity for a century or more. Many of the innovations and locations in this book examine aspects of an industry – including silk, flax and linen – that accounted for as much as a tenth of a countrywide workforce. It was the creation of improved manufacturing capabilities in the textile industry that led the way in terms of power sources, precision machinery, transportation and, of course, factories – with all the social elements that went with them.

Finally, touching on the themes of the Industrial Revolution, it would have been impossible to produce a book like this without covering transport and communication: canals, railways, locomotives, bridges, tunnels – the 18th and 19th centuries saw a huge change in the speed of travel and communication, not just in Britain but worldwide. Much as Britons may have wanted to keep technological developments and industrial pre-eminence to themselves, improved communications meant

▲ *The earliest Bessemer Convertors turned pig iron into steel in a fraction of the time (and cost) of manual labour. From its invention in the mid-19th century, steel was made in sufficient quantities to meet the demand for the explosion of railway building.* Library of Congress

◄ *Before mechanisation, home workers hand-spun and worked yarn to earn a meagre living.* nypl.digitalcollections

▶ *An early 19th-century illustration of Jamaican slaves filling a bag of cotton. The invention of the cotton gin led to the growth of cotton plantations and the consequent increased demand for slave labour to work the fields.*
Library of Congress

▼ *The growth of factories such as these steelworks in Sheffield attracted people from the countryside into towns and cities and promoted the largest ever shift in population in Britain.*
Library of Congress

In many ways the definition of what we know as the Industrial Revolution is that it was a period when science and reason came together to allow people to make mechanisms and structures hitherto impossible both possible and practical. The Scientific Revolution, although based on empiricism, was primarily abstract – as was the Enlightenment's concentration on philosophy and thought. The Industrial Revolution may be so named because before it Britain was an agrarian society and most people lived on the land; after it the trend was towards living in cities and working in dark satanic mills. However, it was primarily concerned with practical things: mechanisation, mass production, manufacturing. You can philosophise about personal transport for as long as you want, but until you can make pedals power a wheel you don't have a bicycle.

So, the starting point and the scope of the book were fairly easy to determine. Where, though, should the book end? A neat end point would have been 1851's Great Exhibition, but I wanted to show the global effects of the subject. What better way to do that than the transatlantic cable and first works on the Suez Canal? Both helped link Britain's Empire and the English-speaking world and contributed materially to the world we live in today. Historians suggest the first flush of the Industrial Revolution was over roughly by 1860, when not only the successes but the failures were apparent: unprofitable industries; railways and canals that were a step too far; innovations and processes that had been overtaken. The Industrial Revolution had also spread – to Belgium (the 'birthplace of the Continental Revolution') based similarly on coal, iron, textiles and transport; to France (although heavily curtailed by political revolution); and to Germany, whose nationalism would foster industrialism. The USA, too, had been pulled along on Britain's coat-tails, but in the later 19th century would begin its growth towards industrial supremacy. The second Industrial Revolution, from the 1860s and 1870s onwards, was more broadly based than the first and, some would say, continues to this day.

Finally, two things are obvious from the moment we start to read about the Industrial Revolution. First, the importance of individuals

that the genies kept jumping out of the bottle and informing the world.

Before the start of the 18th century, most scientific discoveries were just that – driven by pure science rather more than practicalities. Indeed, many have argued that science had little to do with the early years of the Industrial Revolution – although advances in chemistry had immediate effects on the textile industry.

JOHN MARTIN & CO. SHEFFIELD ENGLAND.

FILE AND STEEL MANUFACTURERS & EXPORTERS OF IRON.
W. BAILEY LANG & CO. SOLE AGENTS FOR AMERICA.

◄ *Shop floor view of the milling section of Whitney Automatic Millers, Cleveland, Ohio. More precise engineering led to improved machine tools and the manufacture of interchangeable parts and thus the growth of factories and the spread of mechanisation.* NARA

▼ *An illustration of Jethro Tull's 1731 seed drill from his book* New Horse Hoeing Husbandry *in which he described his agricultural system and his new seed drill. However, it took a century for his ideas to become widely accepted, but when they did they started an agricultural revolution.* Wikicommons

and wealth – the Industrial Revolution was about money. It was driven by wealthy men, or those of wealthy stock, or those who wanted to create wealth. Some of them – such as Titus Salt, John Wilkinson and so on – may have attempted to look after their workers, but most were happy just to accumulate wealth. And the investors who did the same were not investing altruistically: they did so for profit.

Second, the men at the sharp end were more than just inventors – although inventive they certainly were. They were practical men who spent lifetimes involved in making things work. James Watt, Maudslay, the Stephensons, the Brunels: these names keep appearing. As examples of this take two pioneers at different times – Jethro Tull (1674–1741) and John 'Iron-Mad' Wilkinson (1728–1808).

Jethro Tull is known for his seed drill and his single-minded interest in farming. He was a gentleman who studied at Oxford University and Gray's Inn and would have had a career in politics or the law had not illness intervened. He became a farmer, like his father, and soon realised how inefficient hand-sowing methods were. He determined to do something about them. First he looked to education; then, he invented a machine to do the job. Neither

were immediately successful. However, after travelling around Europe, he came home and perfected his machine. In 1731, he published a book, *The New Horse Hoeing Husbandry*, explaining it all. Controversial and not universally accepted, in the years that came Tull's ideas were vindicated and his machinery became commonplace.

John Wilkinson came from a non-conformist

▲ *The Great Exhibition of the Works of Industry of All Nations was held in Hyde Park, London, in 1851. It showcased many of the latest developments of the Industrial Revolution.* WikiCommons

family. His sister married the scientist and political theorist Joseph Priestley. After his apprenticeship with a Liverpool merchant, in 1755 Wilkinson joined his father, Isaac, at the Bersham foundry, near Wrexham in Wales. In 1757, as junior partner and technical manager, he helped erect a blast furnace at Willey in Shropshire. His career makes fascinating reading and when he died he left around £9 million in today's money. (His family then squabbled over the will in a Dickensian manner and most of the cash went to the lawyers.) I think Wilkinson sums up the Industrial Revolution – and the fate of his money sums

up Victorian society! His major achievements, activities and endeavours are as follows:

- In 1768 Wilkinson built a more effective oven to produce coke.
- While most of his life was spent selling iron goods (hence the soubriquet 'Iron-Mad' – he was buried in an iron coffin and his grave was marked by an iron obelisk), he bought shares in Cornish copper mines when the Royal Navy decided that its ships should be copper-bottomed to prevent fouling. His association with Thomas Williams, the 'Copper King', saw Wilkinson buy a share in the Mona Mine at Parys Mountain and shares in Williams' industries. Wilkinson and Williams were among the first to issue trade tokens and in 1785 started the Cornish Metal Company to sell copper. To service his trade tokens, Wilkinson had partnerships with banks in Birmingham, Bilston, Bradley, Brymbo and Shrewsbury.
- He bought lead mines at Minera in Wrexham, and used steam pumping engines to clear them of water. His lead pipe works was at Rotherhithe, London.
- In 1774 he patented a new method of casting and boring cannons – and, later, the cylinders of steam engines.
- In 1775 he was instrumental in pushing through the construction of the famous Iron Bridge in Shropshire before selling his shares to Abraham Darby III in 1777.

▶ *In 1853 New York hosted an Exhibition of the Industry of All Nations, partly to show off the latest inventions and ideas from around the world, but also to proclaim their national pride as an emerging industrial nation. Well over a million visitors attended.* Library of Congress

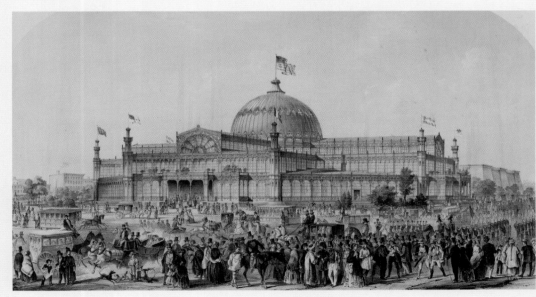

◄ *The Forest River Lead Company was founded in 1840 to manufacture sheet lead.* Library of Congress

- In 1776 James Watt stayed at Wilkinson's house, while emplacing his second steam engine in Wilkinson's works in Broseley.
- In 1787 he constructed the first iron barge in Broseley. Barges took his bar iron by canal to forges in the Birmingham area.
- From 1778 he made iron pipes for the Paris waterworks.
- By 1796 Wilkinson was producing about one-eighth of Britain's cast iron.

▼ *Portrait of John 'Iron-Mad' Wilkinson. By the time he died at the age of 80 he had built bridges and invested in canals and copper mines, founded banks and minted his own coins. He knew everyone who counted, including Joseph Banks and Benjamin Franklin, and perhaps more importantly, supported many other innovators.* Wrexham Museum and Archives Service

▼ *'Dr Jenner performing his first vaccination, on James Phipps, a boy of 8. May 14 1796.' Industrialisation brought people together in living conditions that encouraged sickness and disease. The once clean air was polluted with toxic chemicals and rivers and streams (often already polluted with raw sewerage) were in many cases made worse with the effluent of manufacturing. Jenner discovered a vaccination for smallpox – one of the most devastating diseases – earning himself the title of 'the father of immunology'.* Wellcome Collection

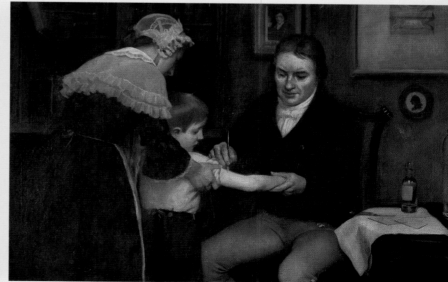

IRON FOUNDRY
ABRAHAM DARBY – SHROPSHIRE, ENGLAND – 1709

Abraham Darby's use of coke for the first time in the smelting of iron ore in 1709 signalled the start of the Industrial Revolution.

By the time of the Industrial Revolution, humans had been producing iron for several millennia – the Iron Age is said to have started around 1200–800BC. However, without a change in production methods to ensure plentiful supplies, the Industrial Revolution

▶ *Abraham Darby's blast furnace at Coalbrookdale as it is now preserved. The upper lintel records the date 1777 – probably the result of its enlargement when it was used in the construction of the Iron Bridge (see page 47). There is some uncertainty about the lower date. It shows now 1638 although there is earlier photographic evidence that shows a date of 1658.* Getty Images

would not have happened at all. Historically pig iron, the crude iron that was the initial product of smelted iron ore, was produced using charcoal, which burns at a higher temperature than wood.

Charcoal – a lightweight form of carbon – was obtained primarily through heating wood in the absence of oxygen. So, the early iron-working industries in Britain tended to be concentrated in areas such as the Weald of Kent, where forests and iron ore cohabited. There were two major problems with the use of charcoal, however. First, demand outstripped supply; second, there were countless other demands upon Britain's limited woodland – most notably from the rapidly growing British mercantile fleet and the Royal Navy.

Something had to happen to allow for an exponential boom in the production of pig iron. Enter at this point Abraham Darby (1678–1717). A Quaker and therefore nonconformist by birth, Darby would have been excluded from many of the professions (this helps to explain why so many of the early industrial and commercial pioneers were Quakers). He had been involved in a brass

▲ *After Darby's furnace ceased production, it gradually disappeared from view and, at one time, there were plans to clear the site. However, it was decided to excavate and preserve the remains and in 1959 – the 250th anniversary of the first use of coke – a small museum opened. In 1970, the Coalbrookdale site became part of the newly established Ironbridge Gorge Museum Trust. The protective building over the old blast furnace was completed in 1981.*
Helen Simonsson/WikiCommons (CC BY-SA 3.0)

foundry in Bristol before he and a group of like-minded entrepreneurs sought to move.

They settled on Coalbrookdale in Shropshire and, in September 1708, Darby took the lease on an existing furnace. The following January – for the first time – pig iron was produced in a furnace using 'charked' coal (coke). The locally produced coal was relatively sulphur free, making it better suited to the smelting process and so establishing the principle of smelting iron without charcoal. From these small beginnings was born the Industrial Revolution.

Today, the remains of Darby's pioneering furnace are on display as part of the Coalbrookdale site of the Ironbridge Gorge Museum Trust – one of a number of museums in an area that, rightly, calls itself 'The Birthplace of Industry'.

2 NEWCOMEN ENGINE
THOMAS NEWCOMEN – CORNWALL, ENGLAND – 1712

The harnessing of steam power was crucial for the efficient pumping-out of mines. Thomas Newcomen's engine of 1712 demonstrated the first practical application of this power.

▼ *A model of one of Newcomen's atmospheric engines. This shows to good effect the domed haystack boiler constructed primarily of copper and lead, as well as the beam that transmitted the power from the piston to the pump. The engine owed much to the work of Thomas Savery and Denis Papin.* Getty Images

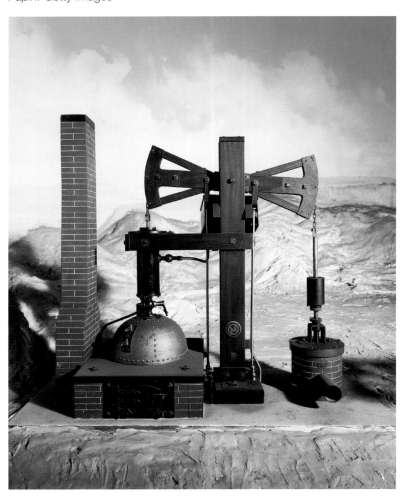

From the late 17th century onwards, a number of pioneers sought to harness the power of steam in engines capable of pumping water from coal and other mines. Water ingress into mines was a serious problem.

Thomas Savery (c1650–1715) was a Devonian engineer who, in 1698, patented what he called the 'Miner's Friend'. He described it as, 'A new invention for raising of water and occasioning motion to all sorts of mill work by the impellent force of fire.' However, it was not an engine as such, as it was incapable of transmitting power to any external equipment. An almost exact contemporary – the Frenchman Denis Papin (1647–1713), who spent much time in Britain – provided some theoretical thoughts on the subject of steam power in his 1690 paper *Nouvelle méthode pour obtenir à bas prix des forces considérables* ('A new method for cheaply obtaining considerable forces').

It was a second engineer from the West Country – Thomas Newcomen (1664–1729) – who combined the practice of Savery and the theory of Papin in the first successful atmospheric, or steam engine. Newcomen's design incorporated a cylinder based on Papin's thought in place of the receiving vessel (where the steam was condensed in Savery's patent design). With the action of the piston, it was possible to work a beam engine. If the other side of the beam engine was linked, via a chain, to a pump at the base of the mineshaft, it was possible to draw out the water from the workings.

Although there is some uncertainty as to the exact location of the first successful Newcomen engine, the earliest documented examples were two erected in the Black Country, of which the earlier is generally accepted as that at the Conygree Coalworks, near Dudley, which commenced operation in 1712. Within three years the first documented engine at work in the Cornish mining industry – at Wheal Vor to

▶ *In 1986 the Black Country Museum in Dudley unveiled a working replica of the world's first successful steam engine, built in 1712 by Thomas Newcomen, which pumped water from coalmines on Lord Dudley's estates.*

the northwest of Helston – was operational. As Newcomen had drawn upon Savery's work, he came to an arrangement with him and operated through Savery's patent.

The importance of Newcomen's development is reflected in the number of machines that he installed – some 100 by the time of his death – and by the widespread adoption of the engine both in the UK and further afield. Newcomen engines were to be found in west and central Europe and possibly even in the early coal industry in Britain's North American territories.

▶ *Newcomen's engine consisted of a boiler (A) situated directly below the cylinder (B). This produced large quantities of very low-pressure steam. The action of the engine was transmitted through a rocking 'great balanced beam', of which the fulcrum (E) rested on the gable wall of the engine house. The pump rods were slung by a chain from one arch-head (F) of the beam; from the second arch-head (D) was suspended a piston (P) working in the cylinder, the top end of which was open to the atmosphere above the piston and the bottom end closed, apart from the short admission pipe connecting the cylinder to the boiler. The piston was surrounded by a seal in the form of a leather ring, but as the cylinder bore was finished by hand and not absolutely true, a layer of water had to be constantly maintained on top of the piston. Installed high up in the engine house was a header water tank (C) fed by a small in-house pump slung from a smaller arch-head. The header tank supplied cold water under pressure via a stand-pipe for condensing the steam in the cylinder with a small branch supplying the cylinder-sealing water; at each top stroke of the piston, excess warm sealing water overflowed down two pipes, one to the in-house well and the other to feed the boiler by gravity. Water and steam egress and ingress were controlled through three valves (V, V^I and V^{II}).*

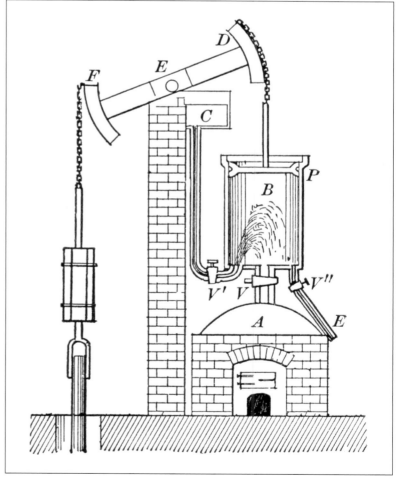

3 WATER-POWERED SILK MILL

JOHN LOMBE – DERBYSHIRE, ENGLAND – 1721

The development of the multi-storey mill was essential if manufacturers were to harness technology in mass production. John Lombe's silk mill at Derby is widely regarded as the pioneering factory.

To William Blake in his poem 'Jerusalem', they were 'dark satanic', but without the development of the multi-storey mill, the boom in the production of cotton and wool, which was one of the hallmarks of the Industrial Revolution, would have been impractical.

As Britain's population increased and as the country became more prosperous, traditional methods of manufacture were no longer capable of dealing with demand. As a result prices rose – and among the affected goods were silk stockings. Traditionally made by framework knitters, hosiery was first manufactured mainly in the southeast. However, in time the industry

▼ *A 17th-century print of John Lombe's silk mill in Derby. Lombe's family was heavily involved in the textile industry: his father was a worsted weaver and his half-brother became a wealthy silk merchant.* Getty Images

moved to the Midlands, where the first factories were constructed. Derby – with its fast-flowing River Derwent and location on the main London to Carlisle road – was an ideal site for initial attempts to build water-powered silk mills. Although the first of these – built for Thomas Cotchett in 1704 – proved a failure, a decade later a mill built for John Lombe (1692–1722) became a success.

In 1716, Norwich-born Lombe travelled to Piedmont, in Italy, to become an employee in a local workshop. At night, by candlelight, he drew copies of the company's silk-throwing equipment and brought his knowledge back to England, where he and his half-brother Thomas (1685–1739) and the engineer George Sorocold (c1668–c1738) constructed a new mill adjacent to Cotchett's failed factory.

Built between 1717 and 1721, the five-storey mill stood upon a series of stone arches that permitted the Derwent in flood to pass beneath the building. Sorocold built a 7m-diameter undershot waterwheel to provide power to the circular spinning machines – known as 'throwing machines'. From the axle of this waterwheel, a vertical shaft took power to each of the spinning floors. The building,

constructed in brick, was 33.5m long and 12m wide. At its commercial peak the site employed some 300 workers and, when completed, the mill represented probably the first fully mechanised factory in the world.

John Lombe took out a patent on the mill's throwing machines, but the King of Sardinia (at the time Sardinia extended far beyond the Mediterranean island into what we now know of as northern mainland Italy) was less than impressed: Lombe had stolen trade secrets in an early example of industrial espionage! The king banned the export of raw silk – and it is also possible that revenge was the cause of Lombe's relatively early death aged only 30 years old.

When the patents expired in 1732, silk manufacture became more widespread in Britain. Lombe's mill continued with silk until it was sold in 1908 to a maker of cough tinctures and other medications. Two years later it was largely destroyed by fire, although surviving fragments were incorporated into a rebuilt structure. After use as a store for the adjacent power station, during the 1970s the building was converted into Derby's Industrial Museum. The building is now known as the Derby Silk Mill.

▲ *Lombe's original mill was to survive for almost two centuries until it was largely destroyed by fire in 1910. The rebuilt mill incorporated elements of the original building and was converted into a museum in the 1970s.* Getty Images

4 ROTHERHAM SWING PLOUGH

JOSEPH FOLJAMBE – YORKSHIRE, ENGLAND – 1730

The Rotherham swing plough saw the first real improvement to the basic plough that was both widely manufactured and commercially successful.

▶ *A contemporary illustration showing the Rotherham plough in use – the original caption reads: 'The Ploughman guides his plough with care and skill.'* Wellcome Collection

▼ *By 1770 the Rotherham swing plough was being used in agriculture across Great Britain and France, as well as in North America. This reproduction is on display in Rotherham museum.* Alamy

Since humankind first took up agriculture, we have needed to plant seeds to grow crops for food. Archaeologists believe that the first rudimentary tools used to break the soil were simply made from appropriately shaped branches. At some stage early humans developed primitive animal or human-drawn ploughs. Although the ancient Egyptians improved these prototypes somewhat, little really changed for several thousand years. When the Dutch began draining their land in the 1600s, they soon realised that they needed better ploughs. As a result they copied Chinese designs, which had iron-tipped, curved mouldboards and adjustable-depth blades.

These adjustments proved to be particularly effective, and when Dutch contractors came to England to drain the East Anglian fens and Somerset moors, it wasn't long before Joseph Foljambe from Rotherham picked up on the designs. He took the ideas further and patented the resulting machine – which was made from cast iron – in 1730. With no depth wheel, it could effectively follow the contours of the terrain, not only giving good results, but requiring fewer animals to pull it. Soon, the new plough became very popular right across Great Britain.

By the 1760s Foljambe was mass-producing the plough in a factory near Rotherham, England – giving the machine its name of the Rotherham swing plough. Probably his most significant achievement was to standardise the interchangeable parts between the ploughs. (Previously, every machine had been a one-off, meaning that no two ploughs could swap common parts, making every one expensive to make.) The resulting efficiency savings significantly reduced costs and made Foljambe's ploughs more affordable to farmers for whom other ploughs had been out of reach. By 1770 many factories had started to produce the Rotherham swing plough and it was in use across Great Britain and France, as well as in North America.

5 FLYING SHUTTLE
JOHN KAY – LANCASHIRE, ENGLAND – 1733

The flying shuttle significantly sped up the weaving process, revolutionising textile manufacture by both reducing manufacturing time and lowering production costs.

▶ *'John Kay of Bury, a Lancastrian Worthy' and inventor of the flying shuttle sitting at his desk.* Getty Images

▼ *Two flying shuttles as designed by John Kay. They have robust iron-tipped ends and rollers on the underside to reduce friction. The lower one has two pins so that it can weave a two-ply thread.* Getty Images

In the 18th century the textile industry was labour-intensive: more or less every procedure was done the slow way, by hand. The need to speed things up was paramount, especially where weaving was concerned. The time it takes for the shuttle (which carries the weft yarns through the warp yarns) to pass across the loom was the main limiting factor. John Kay came up with a device that he called a 'wheeled shuttle'. This was able to travel faster and across greater loom widths, resulting in much faster production. Operable by just one person, it also cut the requirement for a second worker at each loom.

John Kay (1704–79) got his first patent for the shuttle in 1733, but he went on to spend two years further refining the design. In the meantime he formed a partnership and began manufacturing the shuttles commercially. Unfortunately, this caused industrial unrest, with the Colchester weavers in particular being so incensed at the perceived risk to their livelihoods that they unsuccessfully petitioned King George II to have the inventions formally banned.

Although Kay referred to his invention as a 'wheeled shuttle', it wasn't long before it became commonly known as the 'flying shuttle', owing to the speed at which it traversed a loom. This came with its own problems, though – the production of thread was far too slow to keep up with the doubling of weaving capacity in the new shuttle, and the resulting demand.

Unfortunately (although not uniquely among inventors), John Kay did not make much money from this work. His problem was that it cost more to fight patent infringement lawsuits than he won in compensation from them. The main issue he faced was that the manufacturers joined together in a syndicate known as the 'Shuttle Club' that paid the costs of anyone brought before the courts. Frustrated by his inability to collect his royalties in Great Britain, he moved to France in an unsuccessful attempt to seek his fortune there.

6 WINNOWING MACHINE
ANDREW RODGER – ROXBURGHSHIRE, SCOTLAND – 1737

For many thousands of years, the process of separating grain from chaff depended entirely on there being a suitable wind.
The invention of the winnowing machine heralded a new era in agriculture.

▼ *A schematic of how a winnowing machine separates the grain from the chaff.*
Flappiefh/WikiCommons

Historically, agricultural workers always used the breeze to separate grain from chaff (and other unwanted debris) in a process called wind winnowing. It has been around for so long that it is even mentioned in the Old Testament. The principle was simplicity itself – once the grain had been threshed from the stalks, it was thrown up in the air. The wind would then separate out the lighter elements, blowing them away, while the heavier grain fell back to the ground. If the wind's speed was too slow or too fast, however, the operation wasn't possible.

In order to circumvent this, a Scottish farmer called Andrew Rodger came up with a solution. Working from an estate in Roxburghshire, in 1737, he developed what he called a 'fanner'. This was a simple hand-cranked machine constructed mostly from wood that created an

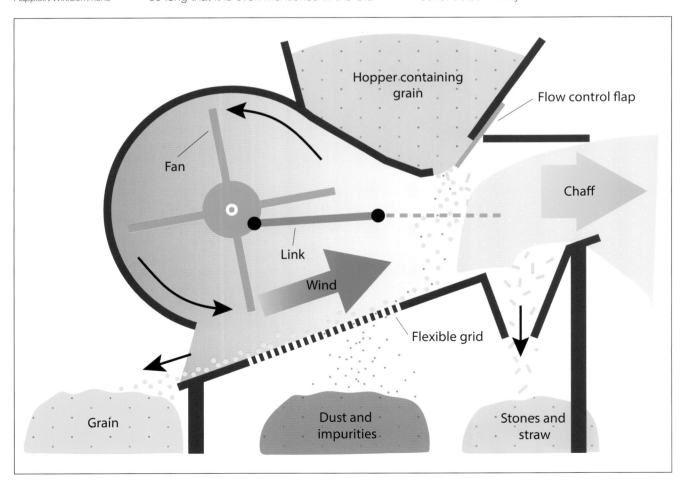

Hopper containing grain

Flow control flap

Fan

Chaff

Link

Wind

Flexible grid

Grain

Dust and impurities

Stones and straw

artificial wind, meaning that farmers could carry out winnowing in any weather conditions.

The operator turned a handle at a steady rate, causing a fan to rotate and creating a blast of air that was directed through a narrow, horizontal duct. Once there was sufficient air speed, a farmer could pour grain into a hopper on the top of the machine, whereupon it fell through the draught that would catch any light debris and blow it out of the other end. The grain itself, being heavier, fell down and through a chute into a collection chamber. It was a simple but effective system that put the choice as to when to do the winnowing back in the farmers' hands.

Some religious ministers considered the wind to be a natural device of God, and so thought the use of fanners was a sin. Despite

▲ *Gustave Courbet's* Les Cribleuses de blé *(Winnowing Wheat) hanging in the Museum of Fine Arts, Nantes, France, depicts three kinds of winnowing. First, on the left there is the most primitive – the languorous state of the woman conducting the hand-sorting shows how tedious and dull the process was. In the middle there is a younger woman undertaking a faster method, but her rigorous posture shows that it is still a task of hard labour. On the right, however, is a winnowing machine being operated by a child, showing the ease provided by mechanisation.* Getty Images

this, Andrew Rodger's family successfully sold the machine for many years. However, as the Industrial Revolution progressed, it slowly lost out to fully mechanised versions.

7

MADELEY WOOD COMPANY
SHROPSHIRE, ENGLAND – 1750s

Madeley Wood works was one of the first to be built specifically to smelt iron with coke. Accordingly, it was one of the earliest ironworks in the world that used coke instead of charcoal to make iron.

Also known as Bedlam Furnaces, the Madeley Wood Company was located beside the north bank of the River Severn, just 1.6km west of Blists Hill in Shropshire, in an early industrial hub of brick and tile works, blast furnaces, and coal, iron and fire clay mines.

In the mid to late 1750s the iron industry in England was experiencing a boom thanks to the demands of the Seven Years' War (1756–63). Manufacturers now appreciated that coke-blast iron was superior to iron made with charcoal, so it made economic sense to build the blast furnaces beside the coalfields. During this period alone, nine coke blast furnaces were built in the Shropshire coalfields.

Established in 1756 in the early days of the Industrial Revolution, the Madeley Wood Company was also known – perhaps even better known – as the 'Bedlam' ironworks, almost certainly because it was adjacent to a Jacobean house called Bedlam Hall and not because of the heat and noise of the works themselves.

Twelve local partners made up the company – none of them from further afield than Bridgenorth, less than 14.5km away. The partners included John Smitheman, the landlord of the site and Lord of the Manor of Madeley.

The Madeley Wood Company built two blast furnaces and brought them into operation in 1757–58. The ironworks held mineral leases in Madeley Parish for the extraction of coal and iron ore – the principal drivers of the Industrial Revolution. The furnaces themselves were among the first specifically designed and constructed buildings for smelting iron with coke, rather than using charcoal. In common with other local works, the company worked the furnaces 24 hours a day. However, depending on the state of the furnaces, they discontinued blowing for a few hours on Sundays, usually between 9 or 10am and 4 or 5pm, allowing the workers a few hours' respite. This was considered to be unusually humanitarian; any longer and the furnaces would cool too much and be out of production for between eight and ten days.

In 1776, the Coalbrookdale Company acquired the Madeley Wood Company. When the resulting organisation was reorganised in 1797, Coalbrookdale was taken over by William Reynolds and Company. By the 1780s almost one third of all the iron worked in England was worked in Shropshire.

▼ *This oil painting by Philippe Jacques de Loutherbourg (1740–1812) has come to epitomise the public image of the Coalbrookdale ironworks. The night-time glow from the flames and the sparks of the furnaces create a true hell on earth. This part of Shropshire rapidly filled with smelting works thanks to the nearby coalfields and plentiful water supply from the River Severn.* Getty Images

8 NEW WILLEY IRONWORKS

SHROPSHIRE, ENGLAND – 1757

The New Willey Ironworks represents an important and innovative industrial complex that saw one of the earliest uses of steam power in industry as well as the building of the world's first iron boat.

▼ *The New Willey Ironworks complex saw one of the earliest uses of steam power in industry – rather than in mining – when introduced in 1774 by John Wilkinson. This drawing shows the ironworks in 1882.* Ralph Pee

Following Abraham Darby's successful smelting of iron without the use of charcoal (see page 12), there was a significant growth in the iron industry in Shropshire. Darby's furnace produced cast iron, a brittle form that needs to be re-melted before it can be converted into more malleable wrought iron or steel. Historically, manufacturers achieved this in a finery forge with an open hearth. During the 18th century, however, there developed new methods of re-melting cast iron. The New Willey Ironworks, south of Broseley in Shropshire, represents one of the most important sites where manufacturers tested and developed new processes and products.

In the mid- to late 18th century, John Wilkinson (1728–1808), who gained the nickname 'Iron-Mad', was one of the most important entrepreneurs in the iron industry. In 1755 he became a partner in the Bersham furnace near Wrexham before, with partners, constructing a blast furnace at Willey two years later. He followed this with the furnace and works at New Willey. In 1768 he built a more effective oven to produce coke. His interests later extended to south Shropshire and south Staffordshire and, in 1774, he invented a technique for the accurate boring of solid cast iron (see page 42).

Opened in 1763, the New Willey Ironworks, located in the Dean Brook Valley, comprised two blast furnaces, innovative casting and boring mills, tramways, and ancillary structures. A water-management scheme, with origins in a 13th-century deer park modernised for industrial use, brought power to the site. It's likely that these works were built with steam power in mind, and in 1776 the site was one of the first to be powered by a commercial Boulton & Watt steam engine (see page 52). The foundry was notable for producing the world's first iron boat (*The Trial*), which was completed in 1787.

Although the bulk of the New Willey site closed in 1804, a small foundry continued in operation through to the 1920s. There are a number of surviving structures – such as the engine house – and many, such as the warehouse and workers' cottages, are now Grade II listed. Remains of further buildings, including the two blast furnaces, are believed to remain buried on the site.

Key

■	Probable route of railway
⋯	Probable route of pack horse track
▬	Road
▬	Dean Brook
1	To further dammed pools
2	To Barrow (today's B4376)
3	To Benthall
4	To Broseley (today's B4376)
5	Rail line down Tarbatch Dingle to Willey Wharf on the River Severn
6	To Dew Corner and Willey
A	Octagonal toll house
B	Today's pond
C	Weigh house
D	Embankment: old dam?
E	Workmen's cottages (part)
F	Large building: stores and offices

Possible slag heap

Furnace

Coke Hearths

9 HYDRAULIC-POWERED BLOWING ENGINE

JOHN WILKINSON – STAFFORDSHIRE, ENGLAND – 1757

The invention of the hydraulic-powered blowing engine made the provision of high-quality iron possible. Without this engine the Industrial Revolution would have seen far fewer advances in technology.

▼ *John Wilkinson's hydraulic-powered blowing engine of 1757 enabled high-quantity, high-quality iron that in turn led to most of the other advances of the era. This portrait of Wilkinson was painted by an unknown artist.* Wolverhampton Arts and Museums Services

The production of quality iron requires high temperatures, which in turn require a plentiful supply of oxygen to the furnace. This is the reason that blacksmiths used bellows to fan their fires. Bellows are all very well for small quantities of material, but for large-scale industrial purposes, the laborious pumping and resulting inconsistent temperatures make them completely inadequate.

An English industrialist called John 'Iron-Mad' Wilkinson (1728–1808) set about finding a solution, patenting a hydraulic-powered blowing engine in 1757. This steam-driven machine was designed to blow vast amounts of air into blast furnaces to make them more efficient. The invention was so successful that Wilkinson was nicknamed the 'Father of the South Staffordshire Iron Industry'.

This pioneering work led to the sudden availability of large amounts of good-quality iron products, in turn enabling a rapid expansion of both the coal and iron industries, which had otherwise begun to wane. Within ten years, Wilkinson had established an ironworks at Bradley, Bilston, Staffordshire – a massive affair that covered 35 hectares, and included not only blast furnaces, rolling mills, forges and so on, but also housing for workers. Other facilities on the site comprised glass works, canal wharfs and a chemical plant.

At that stage most of the manufacturing output was in the form of cast iron. Although cast iron was an enormous improvement over what went before, it is relatively brittle. Consequently, engineers need to take great care in the design of structures made from it, or catastrophe can occur. This led Wilkinson a decade later to start mass-producing wrought iron (which has much better engineering properties) on the site. As a result all manner of the icons of the Industrial Revolution, which had until then been impossible to build, became feasible.

10 MIDDLETON RAILWAY
CHARLES BRANDLING – LEEDS, ENGLAND – 1758

The Middleton Railway was the first railway to be built under instructions from an Act of Parliament. It was also the first railway commercially to use steam traction.

▼ *Published in 1814, The Costumes of Yorkshire included engravings portraying typical work dress in the early 19th century. The background of this image of a collier shows the steam engine* Salamanca *on the Middleton Railway. The toothed driving wheel worked with toothed rail to improve adhesion on steep gradients.* New York Public Library Digital Collection

The first railways predated the Industrial Revolution. In Britain, crude wagonways first appeared in the late 16th century. One notable example – from the 1590s – built at Prescot, near Liverpool, on behalf of local coal owner Philip Layton, transported coal from Prescot Hall for a distance of about 1.6km. In 1603–04, the partnership of Huntingdon Beaumont, lessee of the Strelley coal pits, and Sir Percival Willoughby, local landowner, built the Wollaton Wagonway in Nottinghamshire, again to move coal.

These early wagonways used crude wooden rails along which horse-drawn wagons with iron-tyre wheels could move significantly greater quantities of coal than was practical using the roads. These relatively short lines were constructed between the colliery and the user or point of transshipment (river or canal)

with leased permission from only one or two landowners. For more ambitious schemes over land owned by a number of proprietors, there needed to be an alternative legal structure.

The Middleton area, to the southwest of Leeds, had been coalmining since the 13th century. By the mid-18th century the then owner of the local pits, Charles Brandling (1733–1802), was at a competitive disadvantage as other mine owners were able to move their coal more easily into Leeds by river. Brandling's agent, Richard Humble, decided to exploit the technology of his native northeast and, in 1754, built the first wagonway to serve Middleton. This initial line was over Brandling's own land and needed no permission. However, a more ambitious scheme followed three years later for which Humble sought an Act of Parliament. Brandling received the Royal Assent to construct the Middleton Railway on 9 June 1758 – the first parliamentary powers to construct a railway ever issued. With the principle having been established, from the early 19th century onwards parliamentary Acts became the route by which most railways (until the passing of the Light Railways Act of 1896) were constructed. An Act gave the promoters of the railway compulsory purchase rights over the land where the railways were to run and also regulated the funding of the new lines themselves.

This was not to be the Middleton Railway's only contribution to the development of Britain as an industrial nation. Early in the 19th century, the colliery's manager, John Blenkinsop (1783–1831), developed a new type of iron track, with a toothed rail (patented 1811), in order to employ steam traction to haul trains of coal up steep gradients. The first locomotive, designed by Matthew Murray (1765–1826) and based upon Trevithick's early *Catch me who can*, entered service in 1812. The use of *Salamanca*, the name for this pioneering two-cylinder locomotive, saw the Middleton become the first railway to use steam traction commercially.

11 REDISCOVERY OF CONCRETE

JOHN SMEATON – DEVON, ENGLAND – 1756–59

The rediscovery of hydraulic lime concrete – a lost Roman method of construction – was pivotal in facilitating the building of structures in the sea.

The principle of using light to warn sailors about dangerous reefs and cliffs and to guide them towards safe havens was well established long before the dawn of the industrial age – witness the lighthouse constructed outside Alexandria in the 3rd century BC, which was regarded as one of the Wonders of the Ancient World. However, the sea is a harsh environment – particularly if you are trying to construct buildings to withstand its force – and it was only towards the end of the 17th century that anyone attempted to build lighthouses offshore.

The world's first open ocean lighthouse was built on the Eddystone Rocks, a low-lying reef

▶ *John Smeaton was a pioneering civil engineer at a time when civil engineering was an unrecognised profession. During his work on the Eddystone lighthouse – in the background of this image – Smeaton was to revolutionise contemporary construction through his rediscovery of the Roman means of creating hydraulic lime concrete.* Author's collection

that lay about 13.5km south of Rame Head on the approaches to Plymouth. Built by Henry Winstanley (1644–1703) and constructed in wood, it became operational in November 1698. Modified after its first winter with stone cladding, the lighthouse survived until the Great Storm of 7 December 1703, when Winstanley and five others died as the building was destroyed.

A light on the reef was now considered essential and, in 1709, a new structure – designed by John Rudyard (1650–c1718) – was completed. With a core of brick and stone, this was much more successful, surviving until its lantern caught fire on 2 December 1755. Clad in timber, the building perished.

Then along came John Smeaton (1724–92), the third engineer to attempt the construction of a permanent lighthouse on Eddystone. Using his studies of oak trees as inspiration, he designed a structure made of granite blocks that dovetailed together for added strength. Standing 18m high with a diameter of 8m at the base but only 5m at the top, the tapered tower represented a major advance in lighthouse design. However, arguably, it was Smeaton's use of hydraulic lime concrete that was more significant.

First used by the Romans, but then long forgotten in the mists of time, this type of concrete could set in many extreme conditions, including underwater. With Smeaton's rediscovery, it became possible to build a wide range of structures – breakwaters, bridge piers, and quays among them – using stone and concrete rather than wood. As a result engineers could undertake massive coastal and river works to build structures that were far more durable than their wooden predecessors.

Of Smeaton's lighthouse, it was not the building itself that led to its downfall, but the rock on which it stood. By the 1870s constant erosion resulted in the structure becoming unstable – it used to wobble in heavy seas – and in 1877, following the completion of the fourth incarnation, the lighthouse was decommissioned.

In 1882 the upper section was dismantled under the supervision of William Douglass (1857–1913), the son of the designer of the fourth structure – James Douglass (1826–98) – and re-erected on Plymouth Hoe, where it remains as a tourist attraction. The lower section of Smeaton's original tower proved too difficult to dismantle and so was left in situ – and it remains there still, as a testament to Smeaton's ingenuity.

◄ *When Smeaton's Eddystone lighthouse was replaced at the end of the 19th century, its inherent strength meant that it was impossible to demolish the entire structure and, almost 150 years after its replacement, the stump of the building is still extant. This early postcard records the current lighthouse alongside the stump of Smeaton's when the former was relatively new.* Library of Congress

12 BRIDGEWATER CANAL AND THE BARTON SWING AQUEDUCT

LANCASHIRE, ENGLAND – 1759–61

The pioneering Bridgewater Canal was the first of the industrial age, built at the start of the canal mania, and featured an aqueduct that was the first in Britain over which it was possible to navigate.

▼ *The construction of the Manchester Ship Canal in the late 19th century resulted in the replacement of the aqueduct with a new swing bridge. This view of Brindley's original aqueduct was taken in 1891 shortly before it was demolished.* WikiCommons

Prior to the railway age, water represented the most efficient way to move raw materials and finished goods from manufacturers to customers. Much of the early industrial development, therefore, was predicated on access to rivers. However, from the mid-18th century onwards, canal building enabled industries to develop in locations ill served by rivers.

Certain artificial waterways – such as the Exeter Canal completed in the 1560s – did pre-date the era of industrialisation, but the Bridgewater Canal was the first of a new era. In the mid-18th century Francis Egerton (1736–1803), the 3rd Duke of Bridgewater, owned mines at Worsley to the northwest of Manchester. He had been forced to rely on moving the extracted coal by either pack horse or Mersey and Irwell Navigation (effectively canalised rivers completed by 1734). Then, a visit to the Canal du Midi in France and the construction of the Sankey Canal in northwest England inspired him to develop his own new waterway. With the assistance of his estate manager John Gilbert (1724–95), and later technical advice provided by the engineer James Brindley (1716–72), Egerton produced plans to build a canal from Worsley to Salford. As well as providing improved transport, the canal offered an efficient means to drain water from the line workings. Powers for the canal's construction were obtained by a 1759 Act of Parliament. Further Acts, permitting the extension of the canal from Salford southwards and from Worsley northwards along with several branches, were obtained in 1760, 1762, 1777 and 1795.

The initial section of the canal opened for traffic in 1761. One of its most significant features was the aqueduct constructed at Barton-upon-Irwell by which the canal passed over the River Irwell. Designed by Brindley, the structure was not without its teething problems. For example, one of its three arches buckled under the weight of water when it was first filled. However, the aqueduct opened on 17 July 1761, the first aqueduct in Britain capable of being navigated over. One of the most important canal structures ever completed in the country – and forerunner to such notable aqueducts as that at Pontcysyllte on the Llangollen Canal opened in 1805 (now the oldest and longest navigable aqueduct in Britain) – the Barton Aqueduct stood for more than a century before being replaced

by a swing bridge following the construction of the Manchester Ship Canal in the late 19th century. The swing bridge – designed by Edward Leader Williams (1828–1910) and built by Andrew Handyside & Co. of Derby – was completed in late 1893 and opened to commercial traffic on 1 January 1894. When required to open, the iron trough – which weighs 1,550 tons and is 100m in length – rotates 90 degrees, with gates holding back the water in the trough and on both sections of the canal.

The Bridgewater Canal – one of the few in Britain never to have been nationalised – remains operational, although the southern section no longer connects into the Manchester

▲ *Both the Bridgewater Canal and the Manchester Ship Canal are still in use. Here the* Daniel Adamson *heads downstream along the Manchester Ship Canal with the two swing bridges – the road bridge closer to the camera and the Bridgewater Canal bridge beyond – open to permit its passage. The steamer was originally built in 1903 for the Shropshire Union Canal & Railway Co. and named the* Ralph Brocklebank. *It could also carry passengers. Acquired in 1921 by the Manchester Ship Canal with her sisters, the steamer was renamed after the company's first chairman in 1936. It remained operational until the mid-1980s and berthed for a period at the Boat Museum at Ellesmere Port. There was a serious risk that this historic vessel might be broken up in 2004 until rescue and restoration came.* Andrew/WikiCommons (CC BY 2.0)

Ship Canal. While it no longer carries coal, the waterway remains popular as a destination for those who navigate the canals for pleasure.

◀ *A watercolour image of James Brindley's aqueduct at Barton by G.F. Yates produced about 1793. It shows the passage of a horse-drawn barge on the Bridgewater Canal as it crosses the River Irwell.* Getty Images

13 | HARRISON MARINE CHRONOMETER

JOHN HARRISON – YORKSHIRE, ENGLAND – 1761

The invention of the marine chronometer was essential to help sailors determine their position at sea. The device allowed an accurate measurement of longitude, the east–west position of an object on Earth.

Safe navigation at sea requires a precise knowledge of location based on longitude and latitude. While sailors had learned to determine the latter through measuring the angle of the sun at noon in the Northern Hemisphere and the angle of Polaris – the North Star – in the Southern, the measurement of longitude was more problematic. In the early 16th century, scientists first proposed the use of a chronometer – or time meter – to help. However, that required a clock that could work reliably at sea. By the end of the 17th century, no one had produced a clock capable of replicating the accuracy of a pendulum clock – by far the most reliable timepiece of the era. At the start of the 18th century, Britain was emerging as an increasingly important world power and, in 1714, the British government announced the Longitude Act should be enacted. In effect, this was a competition that, based on various criteria, would financially reward the inventor who came up with the first marine chronometer.

For Yorkshire carpenter and clockmaker John Harrison (1693–1776), winning the prize became a near obsession. He completed his first working example – designated H1 – in 1730. Harrison demonstrated his invention to members of the Royal Society and the clock underwent a sea trial on board HMS *Centurion*. Although not working well on the outbound journey to Lisbon, it proved more successful heading back. Receiving an initial £500 from

◀ *A self-taught carpenter and clockmaker, John Harrison was born in the West Riding of Yorkshire. Following many years' work, he eventually solved the problem of calculating longitude while at sea. His marine chronometer evolved through a number of versions before he finally achieved success in the early 1770s.* Getty Images

▶ *John Harrison developed his second sea watch – now known as H5 – while H4 was undergoing tests. Frustrated by the Board of Longitude, which held back from acknowledging his achievement, Harrison sought an audience with King George III. The monarch was sympathetic and, in 1772, tested the watch himself. He found that over a ten-week period it was accurate to one third of a second per day.* Racklever/English Wikipedia (CC BY-SA 3.0) BUT Science Museum

the Board of Longitude, Harrison undertook further work, resulting in a second model (H2) that he completed in 1741. However, while H2 was an improvement on H1, Harrison recognised flaws in both prototypes. He spent the following 17 years attempting to fix them. Eventually, he produced a third version (H3), which was an improvement on the earlier models but still not wholly accurate.

Parallel experiments to develop watches provided Harrison with the final piece of the puzzle. The 'sea watch' (H4) was first tested on a trip to Jamaica on board HMS *Deptford* that left Portsmouth in November 1761.

Although the return trip proved the efficacy of the watch, the Board of Longitude remained unconvinced and demanded a further test. It was not until the early 1770s, and the direct involvement of King George III, that Harrison's success in the development of the first marine chronometer – and his role in vastly improving maritime safety – was properly recognised. Even then, though, he was denied the full £20,000 prize.

◀ *Today most navigation depends on satellites, but celestial navigation is still a fundamental requirement for many senior merchant marine positions. Modern marine navigation instruments are, therefore, similar to those of earlier centuries: a sextant, navigation map, binoculars (or telescope) and, of course, a chronometer.*

14 THORP MILL

RALPH TAYLOR – LANCASHIRE, ENGLAND – 1764

Much of the Industrial Revolution was based around textile manufacture. Thorp Mill was probably the first water-powered cotton mill in Lancashire and heralded a massive commercial and cultural change.

Although textile manufacture was one of the biggest commercial activities in Great Britain in the mid-1700s, it was still very much a cottage industry. Each part of the process was done slowly and inefficiently by hand. Ralph Taylor was the first to harness water power to drive a cotton mill, which he built by converting three cottages located at Thorp Clough, Royton, Lancashire, in 1764. Sited on a tributary of the River Irk, the whole affair was set up as a carding mill and driven by a waterwheel. Although it ran for only 24 years before closing, the mill showed that there were big gains in efficiency when using water power. As a result, many more water-driven mills followed.

Lancashire was well suited to the manufacture of textiles for several reasons. The most important was that it is close to ports such as Liverpool – making both the import of raw materials and the export of finished goods straightforward. Then, there was a large labour force available that was simultaneously clever and industrious – and accustomed to low agricultural wages. The extra ingredient – and one that is often overlooked – is that the air in Lancashire was predominantly damp. (Dry air causes fibres to snap, so the manufacture of textiles requires a good level of humidity.)

As a result of all these factors combining in one area, the Lancashire textile industry expanded dramatically over the course of a century. This had the dual effect of draining the countryside of its inhabitants and massively increasing the density of local towns and cities as people flocked to the new factories. For example, Royton – the site of Thorp Mill – increased in population more than ten times between 1714 and 1810.

The urbanisation of what had previously been small conurbations was unprecedented and unplanned. This led to all manner of problems, such as rampant overcrowding in low-grade housing and a lack of clean water. Contaminated water led to cholera epidemics in London in 1831–32, 1848–49, 1854 and 1867. Diseases such as smallpox, typhoid, typhus and tuberculosis were also widespread.

▼ *This painting of the town of Oldham – as seen from the village of Glodwick in 1831 – shows the beginnings of the industrial sprawl that resulted from the boom in textile manufacture. The industry walked hand in hand with the Industrial Revolution.* Gallery Oldham

15 SPINNING JENNY
JAMES HARGREAVES – LANCASHIRE, ENGLAND – 1764

The invention of the flying shuttle had doubled the country's weaving capacity but old hand-spun methods couldn't supply sufficient thread. Thankfully, the spinning jenny provided a solution.

▼ *The spinning jenny made a massive contribution to Great Britain's textile industry after the invention of the flying shuttle (see page 19) had doubled the country's weaving capacity. The old hand-spun methods of producing the necessary threads simply couldn't keep up with demand and there weren't enough threads being made to fully exploit the developments in weaving. The spinning jenny, however, radically mechanised thread production, rapidly increasing output.* Getty Images

Although the weaving part of the textile industry had received a huge boost when John Kay invented the flying shuttle in 1733 (see page 19), it was still held back by the fact that there simply wasn't a means to produce enough thread to satisfy the voracious needs of the looms. At that time individual people in their own homes were making threads – a form of manufacture that was both slow and inefficient. It took three decades for technology to catch up, but finally, in 1764, James Hargreaves (1720–78) invented the spinning jenny. This was a multi-spindle spinning frame that he developed in Stanhill, Oswaldtwistle, near Blackburn in Lancashire.

The first spinning jenny had eight spools. By turning a wheel with the right hand, the operator could rotate all of these spools at the same time. Meanwhile, the operator's left hand held a light beam onto which eight rovings (the raw material from which we spin thread) were clamped. These fed onto the spindles. Before long, as Hargreaves improved the machine, the spinning jenny went from eight spools to 80, and then to 120. The thread it produced was coarse and not very strong, but it was still appropriate for several uses. A mere three years later, Richard Arkwright improved both the quality and strength of the yarn itself when he invented the 'water frame' cotton-spinning machine, which made thread suitable for a far wider range of purposes (see page 36).

Britain had been dependent on the woollen trade for centuries and vested interests led to the Calico Acts to restrict imports of various cotton and silk textiles (1700), and to ban the wearing of certain types of cotton cloth (1721). Subsequent pressure by cotton manufacturers saw modifications to these acts (for example the Manchester Act of 1736) before repeal in 1774 paved the way for a massive increase in cotton manufacture, which was considerably assisted by Hargreaves' invention.

16 SOHO MANUFACTORY
MATTHEW BOULTON – BIRMINGHAM, ENGLAND – 1766

Boulton and Watt were among the leading pioneers of the industrial age and their Soho Manufactory was at the forefront of mass production: it was the first true factory in the world.

One of the most significant facets of the process of industrialisation was the adoption of mass production and it was at the Soho Manufactory, 3km from the centre of Birmingham, that these principles were first applied in a factory environment.

Matthew Boulton (1728–1809) was the son of a manufacturer of metal 'toys'. These were items such as snuff boxes, buttons and shoe buckles made out of metal that were ideal for mass production with the appropriate equipment. Boulton was also the first manufacturer of Sheffield Plate – the combination of silver and copper developed in Sheffield following its discovery in 1743 – outside Yorkshire.

In 1762 Boulton went into partnership with John Fothergill (1730–82) having leased a site at Handsworth Heath that was already occupied by a water-driven, metal-rolling mill, and cottages. Under instruction, the Wyatt family of Lichfield demolished the former and replaced it with a new manufactory, completed in 1766. They then demolished the cottages and replaced them with Soho House, which was to become Boulton's home. (It survives to this day and is now a museum.)

Boulton was one of the founders of the Lunar Society – one of the most important intellectual groups of the period with members including Erasmus Darwin, Josiah Wedgwood, Joseph Priestley and William Murdoch – and the group met regularly at Soho.

In 1775 Boulton entered into a partnership with another Lunar Society member, James Watt (1736–1819), primarily with the intention of manufacturing steam engines for the mining industry. Boulton himself had experimented with the construction of a steam engine in 1767, based on the principles of Thomas Savery (c1650–1712), but joined forces with Watt when the latter's original backer John Roebuck (1718–94) became bankrupt. Recognising that if Watt could produce rotative engines, the potential market for steam engines could be extended to other trades, such as the cotton industry, Boulton encouraged Watt to complete his development of such a machine.

The Watt steam engine – which incorporated for the first time a separate condenser – was developed between 1763 and 1775 and the first prototype completed at Soho in 1775. The first examples ran in

▶ *Matthew Boulton was one of the most significant entrepreneurs of the latter half of the 18th century. Born in Birmingham, he established a partnership with James Watt that was crucial in the production of large numbers of steam locomotives. It was these machines that helped power many of the burgeoning mills that were a hallmark of the Industrial Revolution.* Author's collection

industry later the same year. Work continued on the development of a rotary action machine, which Watt finally achieved in 1781 via the adoption of 'sun and planet' gearing (Boulton & Watt used this until the expiry of a patent covering the use of a crank system that had been registered by James Pickard).

Another development on the Soho site was the creation of the Soho Mint. In 1786 Boulton manufactured 100 tons of copper coins for the East India Company. Two years later, he proposed the mass production of low-value copper coinage by machine. Although initially no government order came, the steam-powered mint was established. Each of the eight striking machines was capable of producing almost 85 coins per hour. They produced medals and coins both for the domestic and export market until the mint closed in 1850.

Boulton's original partnership with Fothergill ceased at the end of 1781 and, in 1796, manufacture of steam engines was transferred to a new plant – the Soho Foundry – which was built about 1.6km from the existing Soho Manufactory. Although much of the foundry site remains in use, the original Soho Manufactory was demolished in 1850 and the site was reappropriated for housing.

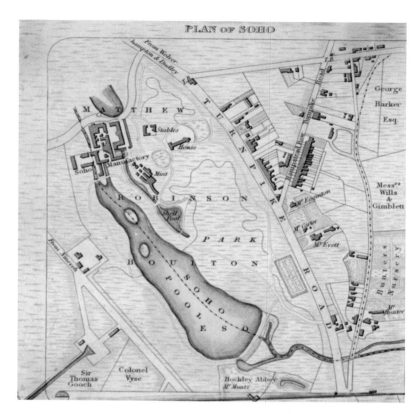

▲ *The Soho Manufactory established by Matthew Boulton and his partner John Fothergill was completed in 1766. Originally designed for 'toy' manufacture – small goods such as belt buckles, buttons and hooks that could be made from metal or other materials – the complex site expanded to include a mint where Boulton pioneered the production of coinage using the most up-to-date methods.* Birmingham Museum

◄ *The design and construction of the Soho Manufactory was undertaken by the Wyatt family of Lichfield. The new building, on a site acquired by a lease in 1761, replaced an earlier water-driven, metal-rolling mill. The building was demolished in 1863, although the house that the Wyatts built for Boulton – Soho House – survives and is now a museum.* Author's collection

17 COTTON-SPINNING MACHINE

RICHARD ARKWRIGHT – NOTTINGHAMSHIRE, ENGLAND – 1767

Arkwright's invention of a multiple-spindle cotton-spinning machine was a significant step forward. The combination of water power with the water frame allowed continuous production using modern employment practices.

Although the spinning jenny (see page 33) had provided the means to produce far more thread than ever before, the thread was coarse and lacked strength. In an attempt to find a solution to these problems, inventor Richard Arkwright (1732–92) joined up with a clockmaker called John Kay to begin development work on a powered spinning frame. The first version was a proof-of-concept machine that spun only four threads at a time.

However, it worked satisfactorily, so Arkwright patented it in 1767. The first large-scale version had 96 spindles, was powered by horses and was installed in a factory built by Arkwright and his partners in Nottingham.

In 1770, Arkwright and his partners switched from using horse power to using water power. They built a dedicated mill on the River Derwent in Cromford, Derbyshire – a factory (or 'manufactory' as it was styled then) that was one of the first of the modern era and implemented all manner of novel industrial practices. These included such things as full employment for workers rather than just temporary contracts. It was also the first factory to have a full throughput process, where raw materials arrived at the gate before undergoing various operations and then emerging as finished products.

▶ *Arkwright and his partners switched from using horses to water power when they built a dedicated mill on the River Derwent at Cromford, Derbyshire in 1771.* Alethe/WikiCommons (CC BY-SA 3.0)

The significance of the water frame is that the use of water power not only provided a faster, stronger and more consistent motion to the spinning process, but also massively reduced the need for skilled labour.

Arkwright's machine, which spun 128 threads at a time, dramatically increased production, but it also greatly improved the quality of the cotton threads, as well as their inherent strength. As specialised craftsmen were no longer needed

▲ *Richard Arkwright's invention of a multiple-spindle, cotton-spinning machine was a significant step towards the era of modern industrialisation. Using water power with the spinning frame allowed for continuous production. This photograph shows a replica cotton-spinning machine.* Moruio/WikiCommons (CC BY-SA 3.0)

to produce the threads, he trained up some unskilled women to run the machines. Before long, mills throughout northern England were using his water-powered invention.

18 CHAMPION'S WET DOCK
WILLIAM CHAMPION – BRISTOL, ENGLAND – 1768

The completion of Champion's wet dock in 1768 was one of the factors that saw Bristol consolidate its position as a key port in Britain and allowed the city to play a significant role in the lucrative – and controversial – slave trade.

Bristol suffers a major disadvantage as a port: it lies some 10km inland. In the 17th century the approach to the harbour along the River Avon was difficult: ships regularly grounded in the tidal river while heading towards the city. In 1662 the corporation

imposed a fine of £10 on any vessels over 60 tons making the passage. Another problem was that sailors beached the ships to permit loading and unloading, potentially causing damage to the wooden-built hulls.

In order to circumvent these problems, in the early 18th century a wet dock – where gates kept the water level constant irrespective of the state of the tide – was completed at Sea Mills in the Avon estuary. This wet dock – based upon an earlier Roman harbour – was not the first to be completed: the Howland Great Dock on the River Thames was constructed between 1695 and 1699 (but lacked unloading facilities) and the world's first commercial wet dock is

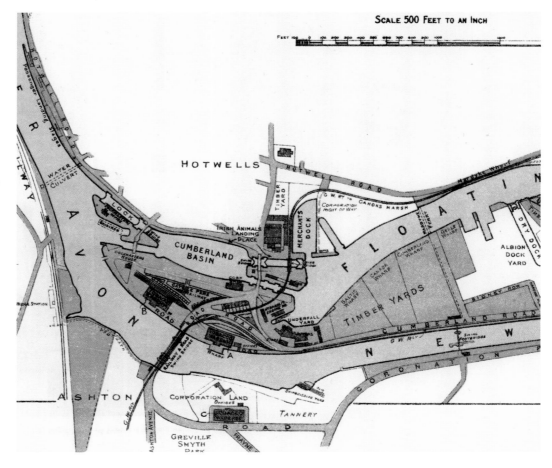

▶ *Champion's wet dock – later renamed Merchants Dock after it was acquired in 1770 by the Merchant Venturers of the City of Bristol – represented the first wet dock established to serve Bristol. Originally, the dock opened directly onto the River Avon, but the development of the docks over the next century saw the route of the river diverted through the New Cut to the south. The original route became a greatly enlarged dock.*
via Peter Waller

generally regarded as the Old Dock at Liverpool, completed in 1715.

Although the wet dock at Sea Mills solved one problem, it created a new one: the road from Sea Mills into Bristol was unsuitable for the movement of goods. This, as well as the fact that Bristol was losing out to Liverpool for the burgeoning transatlantic and colonial trade, meant that the city needed to create improvements to the harbour within the city itself. The first of these was the construction of a privately financed wet dock by William Champion (1709–89) on the north bank of the Avon. Completed in 1768, Champion's wet dock was not an immediate success financially and it was sold two years later to the Merchant Venturers. This guild held a Royal Charter (dating from 1552) that gave it a monopoly

▲ Viewed from the west, Champion's wet dock can be seen immediately to the east of Cumberland Basin. This was opened on 1 May 1809 and was part of the work designed by William Jessop (1745–1814) to counter the competitive edge that Liverpool had gained. via Peter Waller

on Bristol's sea trade. The completion of the new dock resulted in the failure of the earlier dock at Sea Mills, but during the 18th century Bristol was to remain one of the country's most important ports, playing a pivotal role in the slave trade and deriving much profit from that controversial business. It also played a significant part in the trade's abolition in the British Empire, which took place between 1807 and 1833.

Unfortunately, today the pioneering wet dock in Bristol is no more, having been infilled.

19 BINGLEY FIVE RISE LOCKS

JOHN LONGBOTHAM – YORKSHIRE, ENGLAND – 1774

The steepest set of locks ever constructed on a canal in the British Isles, Bingley Five Rise was essential to the development of the national canal network and opened up trade across the Pennines.

▼ *Viewed from the top, the scale of the Five Rise Locks at Bingley is demonstrated to good effect. With an average gradient of one in five, the locks are the steepest completed on any canal in the British Isles and, more than 200 years after their completion, remain in use as an essential part of this important trans-Pennine canal.* Jonathan Forty

In the mid-18th century canals presented an ideal means to improve transport links for industry. They were best built where they could follow the contours of the landscape to avoid the need for additional major engineering works, such as tunnels and aqueducts. However, the commercial pressure to construct canals across the backbone of England – the Pennines – flew in the face of such obstacles, and major work ensued nonetheless.

The impetus for the construction of the Leeds & Liverpool Canal allied with its short branch to Bradford came from several directions. The woollen manufacturers of

Bradford were keen to see their products exported from Liverpool, while the entrepreneurs of Liverpool were eager to see reductions in the price of coal. Powers to permit the construction of the Leeds & Liverpool Canal were obtained in 1770 and James Brindley (1716–72), who had advised on alternative routes, was appointed engineer. However, his early death meant that the clerk of works, John Longbotham (died 1801), took over as chief engineer until his resignation in 1775. He had been involved in surveying the proposed route prior to the Act that permitted construction across the land being obtained.

The section of the canal from Shipley to Skipton along with the branch to Bradford opened in 1774 and 1775. The following year the canal was extended westwards to Gargrave and a year later the link from Shipley to Leeds, and a connection with the Aire & Calder Navigation, was completed. West of the Pennines, the section from Liverpool to Wigan opened in 1781, but work then largely ceased and it was not until 1816 that engineers completed the full route.

The canal eventually stretched some 204km and included 91 locks, eight of which were at Bingley: Three Rise and Five Rise Locks.

The Bingley Five Rise Locks comprises five chambers and six sets of double gates. The locks were – and remain – the steepest set of locks on any canal in the British Isles, with an average gradient of one in five. Designed by Longbotham, the locks were built by locally based stonemasons (including John Sugden from Wilsden, and Jonathan Farrar, Barnabus Morvill and William Wild from Bingley). Over some 98m barges rise or fall some 18m and, when opened on 21 March 1774, the first barge to descend took 28 minutes. The event was recorded in *The Leeds Intelligencer*: 'This joyful and much wished-for event was welcomed with the ringing of Bingley bells, a band of music, the firing of guns by the neighbouring Militia, the shouts of spectators, and all the marks of satisfaction that so important an acquisition merits.' In all, a crowd of some 30,000 people celebrated the opening of the first section of the new canal.

The Bingley Five Rise Locks – now Grade I listed – remain in use, with the Leeds & Liverpool Canal being popular for leisure cruising. The complexity of the use of staircase locks, however, means that a full-time lock keeper is still employed and the gates are locked out of use at other times.

▲ *Recorded in the early 1950s, this view of the Five Rise Locks at Bingley sees a barge emerge from the lowest lock prior to heading east towards Shipley and Leeds.* Julian Thompson/ Online Transport Archive

20 BORING MACHINE
JOHN WILKINSON – STAFFORDSHIRE, ENGLAND – 1774

Steam engines were the primary power source of the Industrial Revolution, but before John Wilkinson invented his boring machine they were very difficult to manufacture to sensible tolerances.

The advance of technology is typically dictated by two things – materials' science and manufacturing capabilities. Nowhere is this better demonstrated than in the development of the steam engine, which was able to produce more power only when it was able to

generate higher pressures. Unfortunately for early pioneers, all the machining methods of the time were rudimentary, and so they could make only low-pressure designs work. For example, James Watt (1736–1819) – famous as an early exponent of steam power (see page 52) – could not find anyone to accurately bore the cylinders for his engine designs. As a result it was impossible to seal the pistons effectively, and they lost significant amounts of energy in leakage.

In 1774, when John Wilkinson (1728–1808) invented a boring machine that was capable of working to far closer tolerances, it provided a major step forwards in industrial technology. Previous designs had used a cutting tool that was held from only one end of the cylinder. Consequently, the forces of the cutting operation caused a lot of tool deflection. This created an uneven surface that was impossible to seal. Wilkinson's version, however, held the cutting tool at both ends of the cylinder – what is known as line-boring. This extra support greatly reduced tool movement, and so significantly improved accuracy. It became possible to seal the pistons far more effectively, allowing them to cope with far higher pressures without losing more energy.

The machine worked so well that Boulton & Watt (see page 52) gave him an exclusive contract to provide cylinders for the company's steam engines.

Having proven that his invention worked extremely well in the cylinder-boring role, Wilkinson went on to widen its application to solve all manner of other engineering problems. Indeed, it is often referred to as the first machine tool in history. The high-pressure steam engines that owed their existence to the creation of the boring machine went on to drive everything from rolling mills and forging hammers in ironworks to the pumps used to empty deep mines in Cornwall.

▼ *John Wilkinson's boring machine of 1774 vastly improved the production of bored components as its inherent rigidity allowed it to work to far closer tolerances. This was especially important in the manufacture of steam engines as it meant they produced more power and at the same time operated more efficiently. Accuracy was also vital in the machining of artillery pieces – as seen here in Jan Verbruggen's drawing of a horizontal boring machine at work at the Royal Arsenal in Woolwich. This is similar to, but not the same as, John Wilkinson's design.* Pieter Verbruggen – Family Archive Semeijns de Vries van Doesburgh/ WikiCommons (CC BY-SA 3.0)

21

THREE MILLS COMPLEX

DANIEL BISSON – LONDON, ENGLAND – 1776

One of the most important industrial heritage sites in England, today only two mills remain at Three Mills, but they are still the largest tidal mills in Great Britain.

▼ *The River Lea Tidal Mill Trust reconstructed the Miller's House (shown on the left) with European Union funding in 1993–94 and it is now Grade I listed. During the Industrial Revolution it provided milled grain for bread and for distilling gin to sustain much of London.* Mervyn Rands/WikiCommons (CC BY-SA 4.0)

The River Lea rises in the Chiltern Hills and flows south and east to become a major tributary of the River Thames, which it joins in east London. Historically, the Lea has been a source of drinking water and water for various industrial uses in the lower reaches around the city.

Three Mills sits in Mill Meads, on an artificial island formed between the tributaries and channels of the River Lea. The waters that power the mills actually flow from Bow Creek, a Lea tributary. The mills were used to grind grain to make flour, but the grain was also used to distil gin, which the mills supplied to the London trade. The mills were even at one time used to make gunpowder.

This area has been known as Three Mills since medieval times and has always been a location for water mills, although little is actually known about them. The Three Mills themselves are some of the earliest extant industrial complexes in London.

Daniel Bisson built the Grade I listed House Mill in 1776 on the site of an earlier mill. Rebuilt once owing to fire damage, it operated until 1941 and remains (almost certainly) the largest tidal mill in the world. Clock Mill and Miller's House were built soon afterwards. By 1878 there were seven water wheels at this location.

House Mill's five storeys are clad with weatherboarding and constructed predominantly from timber, but also from brick on the south side. The building straddles the mill race on cast-iron beams. It has one low breast shot and three undershot water wheels. Powered by impounding the tide to create a 57-acre pond, the wheels drove eight pairs of millstones in one row and four pairs in another. For centuries it milled the grain that was used to make the bread that fed Londoners and for the gin that kept them going. This changed forever during the Blitz in the 1940s, but the building has now been partially restored and fundraising is ongoing to complete the restoration.

Miller's House was almost completely destroyed during the Blitz and demolished in the late 1950s. Clock Mill – of which only the elaborate clock tower is original – was rebuilt in 1817 from an earlier mill using London-stock bricks and was topped with a slate roof. A working distillery, it contained three iron undershot water wheels made by Fawcett & Co. of Liverpool, one 5.9m in diameter and the other two 6.1m in diameter. This drove six pairs of millstones at 130 revolutions a minute and was working until the distillery closed in 1952.

22 | THE WEALTH OF NATIONS
ADAM SMITH – SCOTLAND/USA – 1776

Through the first comprehensive analysis of political economy, Adam Smith kick-started our traditional understanding of economics, identifying the crucial roles of the division of labour, the use of capital and free trade.

Adam Smith (1723–90) – one of the leading lights of the 'Scottish Enlightenment' – provided the economic theory of *laissez faire* capitalism through his seminal work known in its full form as *An Inquiry into the Nature and Causes of the Wealth of Nations*.

Eighteenth-century Edinburgh – the so-called Athens of the North – was one of the most significant places for European economic and political thought, closely rivalled as an academic centre by Glasgow. The Scottish Enlightenment saw writers and academics such as Thomas Reid (1710–96), David Hume (1711–76), Dugald Stewart (1753–1828) and Thomas Brown (1778–1820) all become influential voices in the developing philosophies of the era. The most influential of these thinkers was, however, Smith himself. More than 200 years after his death, his presence as an economic theorist is still highly influential (think of the Adam Smith Institute).

Smith was born in Kirkcaldy, Fife, but was to study under another of the Scottish Enlightenment figures, Francis Hutcheson (1694–1746), at the University of Glasgow. He completed his studies at Oxford (which he regarded as less intellectually stimulating than Glasgow) before working in both Edinburgh and Glasgow. He published his first book, *The Theory of Moral Sentiments*, in 1759, but it was in 1776 that the book for which he will always be remembered – the title of which is commonly shortened to *The Wealth of Nations* – was published. In this and subsequent works, he overturned half a millennium of economic orthodoxy.

Since the Renaissance, the theory of mercantilism – a belief that there was effectively a finite amount of wealth in the world and that economies or countries came to dominate by accruing an ever-greater proportion of that wealth for themselves – had dominated economic thought. It was believed that there were various means by which countries could increase their wealth – conquest, restrictive trade practices, and so on. However, Smith rejected this concept, pioneering the notion of free trade in which private enterprise and profit were the driving forces for increased prosperity. According to Smith's hypothesis, not only was wealth not finite, but through processes such as

▶ *Adam Smith was one of the key figures of the Scottish Enlightenment in the mid- to late 18th century. This group of thinkers, which also included individuals like David Hume, was highly influential in creating the philosophical and economic theories that underlay the Industrial Revolution.*
Library of Congress

► *The frontispiece to the first volume of the second edition of* An Inquiry into the Nature and Causes of the Wealth of Nations. *The second edition was published in 1778. Adam Smith's seminal work provided a cornerstone in the theory of the free-market economy and has continued to influence economic policies through to the modern age.* Author's collection

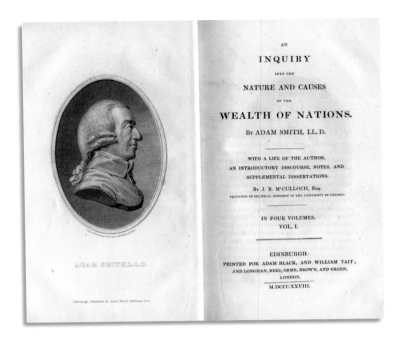

the division of labour and competition, we were able to create it.

At the start of the 19th century, most economies were still protectionist but as Smith's theories became more widely accepted, so the concept of free trade and *laissez faire* became prevalent. There was controversy – witness the debates over the abolition of the Corn Laws in Britain during the 1840s – but by the end of the 19th century, free trade and competition were dominant in most of the world's major economies.

▼ *Adam Smith died in Edinburgh aged 67. There's a monument to him outside St Giles' Cathedral. The bronze was created by Alexander Stoddard and unveiled in 2008.*

23 SPINNING MULE
SAMUEL CROMPTON – LANCASHIRE, ENGLAND – 1779

The spinning mule was yet another step forward in the production of yarn for the textile industry. At its peak there were 50 million mule spindles running in Lancashire alone.

▼ *The spinning mule – a small version of which is shown here – was another step forward in the mass production of the yarns used in textile manufacture. Each machine could have up to 1,320 spindles and be 46m long. At the peak of its use, there were 50 million mule spindles running in Lancashire alone.* Bolton Library and Museum Services

Up until the 1770s, the production of textiles had been a cottage industry in which women manufactured yarn, spinning it from rough rovings. This was a slow, labour-intensive process, and when John Kay introduced the flying shuttle (see page 19), the women simply could not make enough yarn to keep up with the weavers' demand. When Samuel Crompton (1753–1827) invented the spinning mule, a machine used to spin cotton and other fibres, in 1779, everything changed.

Essentially a clever combination of Arkwright's water frame (see page 36) and James Hargreaves' spinning jenny (see page 33), Crompton's first wooden mule had 48 spindles and was able to manufacture 0.45kg of a strong, thin thread every day. Suitable for use as both warp and weft yarn, the mule was soon in great demand. The machine itself still required a significant amount of labour – each machine needed a minder, as well as two boys, known as the little piecer and the big or side piecer.

Crompton could not afford to patent his design, so he sold it to David Dale, who went on to profit from it commercially. Various people then got involved with further developing the machine, making a series of small refinements to both the mechanisms and the construction materials. For example, the drive belts were soon replaced with gear trains and the rollers changed from wood to metal. The first mules were animal-powered, but as time went on they increasingly became powered by water.

Although the Crompton design was superseded by Richard Roberts' self-acting mule of 1825 (see page 111), the basic principles did not change very much, and mules were the most widely used spinning machines from 1790 right up until the end of the 19th century by when a typical cotton mill would have more than 60 mules, each with 1,320 spindles. Such machines were still used for manufacturing fine yarns until the early 1980s.

▼ *An engraving of a thoughtful Samuel Crompton by James Morrison, after a painting by Charles Allingham in c1800.* Author's collection

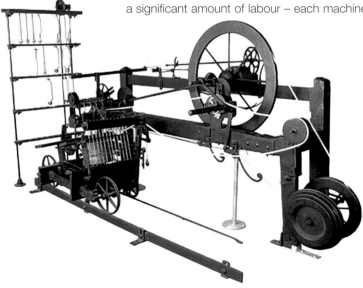

THE IRON BRIDGE

ABRAHAM DARBY III – SHROPSHIRE, ENGLAND – 1779

24

The Iron Bridge in Shropshire was the first major bridge to be constructed in cast iron. Its success demonstrated the huge potential for this new material.

The success of Abraham Darby in smelting iron from coke in Coalbrookdale (see page 12) resulted in the rapid economic development of this part of Shropshire. As the economy grew, so there was recognition that the existing river crossings – reliant upon the bridge at Buildwas more than 3km upstream – were inadequate.

In planning for a new bridge between Broseley and Madeley, there were a number of factors that determined the nature of the structure. The river was still the main artery carrying traffic to and from Shrewsbury, so the bridge required a single span that would allow sufficient clearance for ships to pass beneath unhindered. Designers also had to take into account the steep sides and instability of the gorge. Architect Thomas Farnolls Prichard (c1723–77) first proposed the concept of an iron bridge in 1773 in a letter to one of the local

▼ *The first major bridge to be constructed in cast iron, the bridge had to be of sufficient height to permit ships to continue to use the navigable River Severn through to Shrewsbury.*

ironmasters, John Wilkinson (1728–1808). Two years later, with Abraham Darby III (1750–89), grandson of the pioneering Abraham Darby, as treasurer, a subscription was raised to construct a single-span bridge. An Act of Parliament in March 1776 granted powers to construct the bridge and Darby was commissioned to cast the requisite iron for the build. Doubts about the construction, however, resulted in the withdrawal of the commission two months later, as the trustees sought to create the bridge using more conventional materials. When no suitable new design emerged, the trustees reverted to the concept of an iron bridge.

Work began on the bridge's foundations in November 1777 with the erection of the stonework on both banks for the two abutments. With these completed, the bridge builders began the ironwork and, on 2 July 1779, they completed the span across the river. The bridge finally opened on 1 January 1781.

The completed structure has a span of 30.63m and was built in a way that would have been recognisable to carpenters building a wooden bridge. It is formed of five parallel ribs that incorporate almost 1,700 separate pieces of cast iron. Each piece was unique and so there are slight discrepancies in size between items that are ostensibly the same. In all, the structure required almost 385 tons of iron, which Darby alone supplied.

Since its opening, the bridge has undergone several significant repairs – partly as a reflection of the gorge's instability with numerous landslides recorded in the local vicinity. (In the early 1970s, engineers constructed a ferro-concrete invert in the river to counter the inward movement of the bridge that would otherwise have led to its failure.) With the growing weight of vehicular traffic, the bridge was pedestrianised in 1934 (the same year in which it was made a Scheduled Ancient Monument). Now a UNESCO World Heritage Site, the bridge remains a major tourist attraction.

25 SHOT TOWER PROCESS
WILLIAM WATTS – BRISTOL, ENGLAND – 1782

Until the invention of the shot tower, the production of lead shot for use in shotguns was a slow, expensive and haphazard affair. The new process revolutionised manufacture, increasing the supply of shot and lowering its price.

▼ *A schematic showing the rationale behind shot tower production. The fire at point B melts the lead at point A. When it is ready, the lead is poured through a sieve at point C. From there it falls to point D, where there is a container of water. While it is falling, surface tension in the lead forms it into near-perfect spheres. The water cools and solidifies the spheres in their shape.* Science Photo Library

Shotguns, which are used for both sport and hunting for food, fire a large number of small, round lead balls at targets that are typically 35–45m away. They were very popular in the late 1700s, but the production of the shot itself was primitive, making it expensive and of variable quality. A plumber called William Watts came up with an idea that he thought might reduce the production costs and improve consistency. He reasoned that if he poured molten lead through a sieve and allowed it to fall into cold water, the result would be spherical shot.

After some experimentation, Watts realised that the molten lead needed to fall a long way in order to have time to cool down and form spheres. In 1775, he began to convert his house in Redcliffe in Bristol to make it suitable for shot production. He added a tower to the top, and dug a shaft below it, giving him an overall drop of 27m. He placed a container of water at the bottom of the drop, then poured some molten lead from the top through a screen made of perforated zinc. As the molten lead fell, surface tension caused it to form lots of small spheres that cooled enough to solidify and land as solid spheres in the cold water below. Watts sorted the balls for size and simply re-melted any that were not of a good enough standard.

The process worked well and Watts was granted a patent on it, but unfortunately his business acumen was not as good as his innovative skill, and within 20 years he was declared bankrupt. The system of manufacture was a good one, though, and across the world many others built towers along the same lines – adding height to give the balls more time to cool when the manufacturer was aiming for larger shot.

The shot tower at Boughton, Chester, which stands alongside the Shropshire Union Canal, was built in 1799, making it the oldest of its kind in the UK and, possibly, the world. The introduction of such structures revolutionised the production of the lead shot used in shotguns and significantly reduced its cost. Rept0n1x/WikiCommons (CC BY-SA 3.0)

26 *PYROSCAPHE*
CLAUDE-FRANÇOIS-DOROTHÉE – LYON, FRANCE – 1783

There had been many attempts to design a steam-driven boat, but the Marquis de Jouffroy d'Abbans' *Pyroscaphe* (fireboat) was the first success and paved the way for a revolution in powered shipping.

▼ *Statue of Marquis de Jouffroy d'Abbans (1751–1832) facing the River Doubs in Helvetia Park in Besançon, France. Many consider him the first inventor of the steamboat. His vessel* Pyroscaphe *was built in 1783. He tried to patent his invention, but his attempt was thwarted by the Académie Française and then the outbreak of the French Revolution.* Arnaud 25/ WikiCommons (CC BY-SA 4.0)

Claude-François-Dorothée, Marquis de Jouffroy d'Abbans (1751–1832), laboured for years experimenting with steam-powered vessels. The Marquis' first (partially) successful attempt was the *Palmipède* powered by a Newcomen engine on the River Doubs at Baum-des-Dames in Franche-Comté in 1776. His next vessel was *Pyroscaphe*, an altogether larger boat that he demonstrated on the Saône at Lyon on 15 July 1783. She was 13m long and her oars were equipped with rotating hinged flaps that were modelled on the webbed feet of wading birds. Thousands of spectators came to watch the trial. The details are vague, but after about 15 minutes' steaming, it seems the hull split under the strain of the pounding engine and the boiler leaked steam. These were typical problems – the boat was repaired and managed a few more sorties.

However, the Marquis was not satisfied and set about improving the *Pyroscaphe*. At the next attempt she had a crew of three, displaced 163 tons of water and was over 46m long with a beam of almost 4.6m. The horizontal engine moved a reciprocating double rack that was geared to ratchet wheels on a shaft that supported a large paddle wheel on either side of the boat. This double-ratchet mechanism powered the continuous rotation of the paddle wheels.

For 16 months the *Pyroscaphe* carried freight and passengers between Lyons and Lile Barbe. Flushed with success, the Marquis applied to the government for a licence to start a steamship company, but the bureaucrats passed the buck to the French Academy of Science for review.

The Marquis tried to patent his invention and made a model for inspection, which is now in the Musée de la Marine in Paris. He wanted to build a full-sized version to test on the Seine in Paris, but he was refused permission. The academicians declared the

◀ Pyroscaphe – *a bateau à vapeur – was first publicly demonstrated in front of a crowd of thousands of excited spectators on 15 July 1783 on the River Saône in France. This illustration appeared in* Le Petit Journal, *23 December 1900.* Art Media/Print Collector/Getty Images

results inconclusive, the government refused the Marquis' request and then the Revolution broke out and the patent was never granted.

The Marquis wrote a treatise on steamships and then stopped experimenting and disappeared into self-exile. Instead, an American inventor, Robert Fulton (see page 58), developed a successful steamboat and is generally credited with finding the solution to steam-powered ships. Magnanimously, Fulton acknowledged his debt to the Marquis, but the Marquis himself died in Paris in 1832, a forgotten and disillusioned man.

By the middle of the 19th century steam-powered ships had completely displaced sailing ships for international trade.

27 WHITBREAD ENGINE
MATTHEW BOULTON AND JAMES WATT – SCOTLAND – 1784

James Watt's invention of the steam engine was a crucial development in the Industrial Revolution. The new steam-driven technology inspired engineers and inventors, who in turn devised new machines and industries that would drive the future of the British economy.

Watt (1736–1819) was a Scottish instrument maker who saw how he could take the current static technology of a steam-driven working cylinder and link it to a separate condenser to create a partial vacuum (also known as an atmospheric engine). His innovative idea, which he patented in 1769, was that the working cylinder would be permanently hot and the condenser cool. He worked on the idea with Matthew Boulton (1728–1809), a Birmingham industrialist, and together they

▼ A 1784 illustration of an early Boulton & Watt steam engine. Their development of steam power put them at the forefront of the Industrial Revolution. It is no exaggeration to say that without their work the technological developments of the 18th and early 19th centuries would not have happened. Author's collection

▼ James Watt – who gives his name to the SI unit of power – was in many ways the father of the Industrial Revolution. His engine, particularly after partnership with Matthew Boulton, would provide the Industrial Revolution with its power.

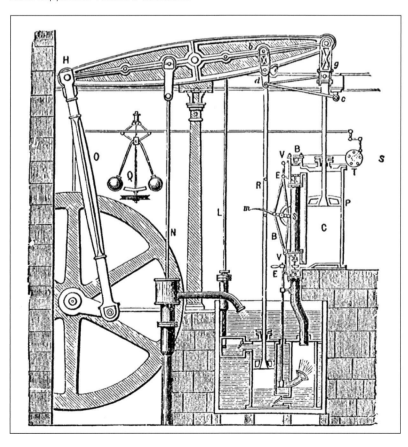

put their design into production in 1775. Watt continued to develop his beam engine and within a few years had sealed the cylinder at both ends. He introduced steam at either end of the piston in the cylinder to create a double-action engine that produced power with both the up and the down strokes.

The engine was built for Samuel Whitbread (1720–96) in 1784 to replace the existing horse-driven mill at his brewery in London. Installed the following year, its added power helped Whitbread become the largest brewer in Britain. The engine's drive gear was connected to a number of wooden line shafts that, in turn, connected to rollers to crush malt, an Archimedes' screw to lift the malt into a hopper, a hoist for lifting heavy items, a three-piston pump to move the beer, and a stirrer inside the vat. Another pump, connected to the engine's

beam, pumped water from a well on the ground up to a tank on the roof. The engine remained working at the brewery until 1887.

The Whitbread Engine was one of the first rotative beam engines in which the reciprocating motion of the beam is converted to rotary motion to produce a continuous power source. Once it was out of service, instead of scrapping the revolutionary machine, Whitbread donated it to the Powerhouse Museum in Sydney, Australia, as an educational tool. It has now been restored to full working order.

▼ *The oldest surviving example of the Boulton & Watt steam engine is known as the Whitbread Engine and is on permanent display in the Powerhouse Museum, Sydney, Australia. It was one of the first rotative engines and uses sun and planet wheels instead of the more common crank to drive the flywheel.* Newtown Grafitti/WikiCommons (CC BY-SA 2.0)

28 LIGHTHOUSE LIGHT
AIMÉ ARGAND AND MATTHEW BOULTON – BIRMINGHAM, ENGLAND – 1784

The Argand lighthouse lamp greatly improved the visibility of lighthouse beams, making them brighter and more reliable. The lamp saved numerous lives at sea and became the industry standard.

▼ *Diagram showing the workings and airflow of the Argand lamp.* Terry Pepper

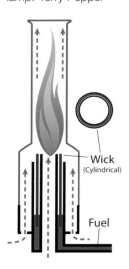

Wick
(Cylindrical)

Fuel

▶ *Diagram of a Fresnel lighthouse lens showing how the light was diffracted by the varying annular sections and angles of the glass. The lens threw out light to far greater distances than previous lenses and provided early warning to mariners of treacherous coasts.* WikiCommons

Before the late 18th century, burning wooden pyres or multiple candles illuminated our lighthouses – with all the associated problems that come with open flames. However, in 1782, the Swiss physicist and chemist Aimé Argand (1750–1803) and his English partner Matthew Boulton (1728–1809) revolutionised lighthouses with a new lamp that became the industry standard for more than a century.

The innovation consisted of a hollow cylindrical wick that allowed more oxygen to reach the flame. This, in turn, produced a brighter, more stable light. He then placed a chimney over the burner and improved the updraft of oxygen to steady the flame even further. This system burned the fuel more efficiently and economically, gave off brighter light, and produced less smoke. The wick was supplied by a gravity feed from a reservoir mounted above the burner (as opposed to wicking the oil from a lower reservoir). Lamps typically used whale oil, colza (rapeseed oil), olive oil or other vegetable-oil fuel. The lamps, however, could be dangerous as they were top heavy. Argand patented his idea in 1780, but did not profit from his work. Others, especially in the USA, exploited his ideas without paying him for them.

In 1784, Argand went to England to find a partner with whom to manufacture his lamps, in particular to produce reliable lamps for lighthouses. He found Matthew Boulton, a Birmingham manufacturer and pioneer in metalwork who also worked with James Watt (see page 52). Early examples of their lamps used ground glass that was occasionally tinted around the wick. Later models used a mantle of thorium dioxide suspended over the flame, which created a bright, steady light.

FRESNEL LENS
Originally developed by the French physicist Augustin-Jean Fresnel (1788–1827), this lens was specifically for use in lighthouses to throw the illumination far out to sea. Instead of using one giant lens ground down to provide different focal lengths, Fresnel devised a multi-lens which is divided into a set of concentric annular sections (it also reduced the amount of glass needed for the lens). The full lens comprised a combination of curved and flat surfaces.

The first documented use of a Fresnel lens was in 1823 for the Cordouan lighthouse at the mouth of the Gironde just down from the important port of Bordeaux in France. The light was visible for a radius of over 32km.

Fresnel produced six different sizes of lens for lighthouses, varying in focal length. Combined specifically with the Argand-Boulton lamp, the lens was able to save thousands of lives at sea.

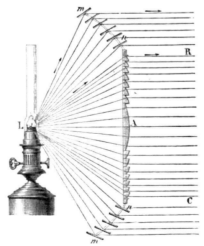

29 POWER LOOM
EDMUND CARTWRIGHT – DONCASTER, ENGLAND – 1785

Edmund Cartwright's power loom revolutionised the textiles industry. Although the design was far from perfect, it featured many significant innovations and paved the way for even better looms later on.

▶ *Edmund Cartwright was both an inventor and a Church of England clergyman. This is an engraving of him by James Thomson, after a painting by Robert Fulton.* Science Photo Library

▼ *Edmund Cartwright invented the power loom, as depicted here. Although the design was far from perfect, it was yet another advancement in the manufacture of textiles. It featured many significant innovations and paved the way for later improved versions.* Getty Images

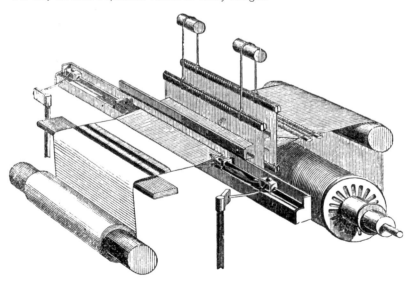

Edmund Cartwright (1743–1823) was fascinated by the conundrum of how to power a loom using something other than the human hand. His first attempt at solving the problem – in 1784 – proved to be worthless. However, his second attempt, patented in 1785, was far better. This established a number of precedents – it included a series of detailed mechanisms that controlled the yarn in specific ways and vastly improved performance. Although these mechanisms were quite crude, they demonstrated the principles well, and all were later improved on by other inventors.

However, the invention was not popular with the hand-loom weavers, whose jobs were threatened. The strength of feeling was so strong that, for example, when industrialist Robert Grimshaw (1757–99) built a weaving factory at Knott Mill, Manchester, in 1790, intending to equip it with 500 of Cartwright's power looms, the factory burned down under suspicious circumstances after only 30 of the looms had been installed.

Undeterred by these issues, Cartwright began manufacturing fabrics in Doncaster using the latest variants of his power loom. As he began production, various problems came to light with his machines and he continually made improvements in order to circumvent issues and perfect the loom. These included tweaks such as a device that stopped the loom if a shuttle failed to enter a shuttle box and the addition of special automatic cloth-stretching 'temples'. Unfortunately, Cartwright was not a very good businessman and creditors repossessed his mill in 1793.

Nonetheless, although his loom design was not a commercial success in its own right, it revolutionised the textiles industry. In gratitude for what he had done for the country's economy, in 1809 parliament awarded Cartwright a grant of £10,000 for his invention, and he was elected a Fellow of the Royal Society in May 1821.

30 THRESHING MACHINE
ANDREW MEIKLE – EAST LOTHIAN, SCOTLAND – 1786

Threshing cereals to separate the husks from grain is a major part of the harvesting process. Andrew Meikle's threshing machine massively boosted grain production and was a significant contributor to the advancement of milling.

▼ *Mechanisation of rural jobs was a significant feature of the Industrial Revolution, including the adoption of Meikle's threshing machine. In the later 19th century, steam power would lead to further innovation.*

Threshing cereals has been a part of human agriculture since harvesting crops first began. The process is mentioned throughout early literature in everything from the Bible to the tales of the ancient Greeks. However, up until the late 1700s, the process for threshing had hardly changed, despite attempts in the early 1700s by an inventor called Michael Menzies (died 1766) to find a solution that would make the process more efficient.

In 1778, millwright Andrew Meikle (1719–1811), who worked out of Houston Mill in East Lothian, Scotland, began developing a new threshing machine. At first he may well have attempted something based on the Menzies design, but it didn't work. Instead, he tried a different method based on a threshing machine that had been created in Northumberland. However, that was also a failure. Nevertheless, Meikle did not give up. Rather, he came up with a system that worked in a completely different way, based on a principle that was similar to the way in which the fibres are separated out of flax plants in a flax-scutching machine.

The new design used a robust drum with beaters fixed to it. These flayed the cereals, removing the outer husks from the actual grains (previous machines had simply rubbed them). He patented the idea in 1788, after which he began manufacturing. One of his customers was George Washington, who ordered a machine for export to the USA in 1792. Washington (who was a farmer as well as a soldier and politician) was excited when his agent told him that Meikle's machine

HUSBANDRY.

Plate X.

'threshes eight quarters [64 bushels] of oats an hour, with four horses and four men'. Unfortunately, Washington's interest was not enough to keep Meikle afloat, however, and it seems that his business was not a commercial success: he was destitute in 1809, and died two years later.

▲ *The threshing machine, as invented by Andrew Meikle, provided a much faster way to separate the husks from the grain when harvesting cereals. Although commercially unsuccessful, the machine was a significant contributor to the advancement of milling, and so helped speed up food supply at a time of increasing populations. This engraving shows a horse-powered thresher* (above) *and the improved form powered by waterwheel* (below). *Wellcome Collection*

31

FIRST REGULAR STEAMBOAT SERVICES

JOHN FITCH – DELAWARE RIVER, USA – 1787

John Fitch introduced the first regular steamboat service and Robert Fulton developed a wider service that led to the rapid transport of goods across the North American continent, fostering the American Industrial Revolution.

John Fitch (1743–1798) invented the first steamboat and began the first regular steamboat service in the USA. His boat featured a line of canoe-like paddles mounted on either side of the vessel to power it through the water. In the summer of 1785, he took the plans to the Continental Congress in the hope of getting investment to help him bring his ideas into reality. Eminent figures such as Benjamin Franklin and George Washington were impressed, but nobody wanted to invest.

So, Fitch built his own steam engine and when eventually he found investment, launched his steamboat in 1787. He ran it along the Delaware River, between Philadelphia and Burlington, but still the authorities themselves were not keen to back him. Instead he built a bigger and faster steamboat with stern-mounted paddles and a more compact water boiler. By 1790 he could offer three round trips a week down the Delaware between Philadelphia, Bristol and Trenton. The fare was cheaper than the rival stagecoach and offered free sausages, rum and beer. However, despite successfully steaming almost 4,850km that summer, the venture was not popular enough to make a profit. Fitch was granted a US patent for his steamboat, but his investors abandoned him. In desperation Fitch travelled to France, but the French were in the midst of revolution and had their minds on other things. After

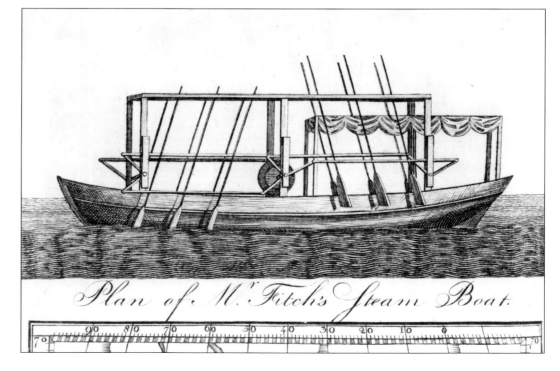

▶ *John Fitch demonstrated the first successful steam-powered boat on the Delaware River in 1787. Named the Perseverance, she contained a wood-burning boiler that powered six sets of vertical wooden oars that moved her through the water at 5km/h.* Library of Congress

Plan of M.r Fitch's Steam Boat.

returning to the USA, he neared destitution and took his own life. He was buried in a pauper's grave.

Robert Fulton (1765–1815) was born in the British colony of Pennsylvania, and started out as a portrait painter, gaining fame in London for a portrait of Benjamin Franklin. He diversified into canals and shipbuilding and was intrigued with the possibilities of the steam engine – especially how it could power ships.

In 1800 Fulton successfully tested the first working submarine ('diving boat') called *Nautilus*. Neither the French nor the British were interested enough to invest in it and on his return to the USA in 1802, Fulton went into partnership with Robert Livingstone and won an exclusive licence for steamboat services along the Hudson River.

Fulton took a special English steam engine and installed it in a flat-bottomed, square-sterned boat called the *Clermont* (previously known as *Fulton's Folly*). She made her first appearance on 17 August 1807 when she steamed 240km from New York to Albany. The trip took 32 hours at an average speed of 8km/h. This vastly quicker river journey made the venture a triumph and a regular service was put into place. Within five years Fulton was hugely successful, running regular steamboat services along six major rivers, including a steamboat and freight service between New Orleans, Louisiana and Natchez, plus another across Chesapeake Bay.

Although not an inventor as such, Fulton was a visionary. His forward-thinking ideas promoted the North American Industrial Revolution by helping manufacturers quickly transport goods and materials across the vast continent. It also opened up the way for exploration and settlement, securing America's economic future.

▲ *Robert Fulton's* Clermont *was the first steam-propelled regular riverboat service. The venture along the Hudson was such a success that he soon offered scheduled runs along stretches of six major rivers. These, in turn, helped to open up the interior of the American continent and contributed to the growth of the Industrial Revolution in North America.* New York Public Library Digital Collection

32 GAS ILLUMINATION
WILLIAM MURDOCH – CORNWALL, ENGLAND – 1790s

Today we take for granted the illumination of our homes, workplaces and cities when it's dark. William Murdoch's innovative use of coal gas made this possible on a large scale for the first time.

William Murdoch (1754–1839) was a Scottish engineer and inventor who moved to Birmingham to work with James Watt (see page 34) at his pattern shop, the Soho Foundry. There, he soon became an expert in fitting and erecting steam engines. In 1779 he was sent to Cornwall to work as a steam-engine erector and maintenance engineer for the Boulton & Watt engines (see page 52) pumping water out of local tin mines.

While in Cornwall Murdoch was responsible for many steam-engine refinements as well as other scientific discoveries. One such discovery was the replacement of expensive Russian isinglass with much cheaper dried British cod extract for clarifying the impurities from beer. This saved so much money that the Committee of London Brewers paid him £2,000 for the right to use it.

▶ The house in Cross Street, Redruth, Cornwall where William Murdoch lived and experimented with gas lighting. This may also have been the first domestic building to have been lit by gas illumination, but Murdoch more probably first lit the Soho Foundry or his own workshop. Tim Green/WikiCommons (CC BY-SA 2.0)

Between 1792 and 1794, at a time when homes and businesses were lit using oil and tallow, Murdoch devised his best-known invention, domestic lighting. One day he was relaxing by the fire when he noticed that when he filled his pipe with coal dust and put it near the flame, a gas escaped through the mouthpiece and, ignited, shone a bright light. He began to experiment with light created by the gas collected from the combustion of coal and other materials. Nobody knows for certain how he did this, but it was probably by filling a small retort with burning coals and collecting the offset gas. He piped this via a long iron tube attached to an old gun barrel and then lit the exhaust fumes to produce a light. It was said that he used such lighting at his cottage in Cross Street, Redruth, or at his home in Soho, but more reliable reports from witnesses to his demonstrations say that the apparatus was in his workshop or foundry.

Murdoch continued to refine his invention, experimenting with a variety of substances to

▲ *The Soho Foundry was built beside the Birmingham canal in 1795 at Smethwick in the West Midlands. It was established by Matthew Boulton and James Watt for the manufacture and production of steam engines. The factory itself was a model of sophisticated planning and management techniques, and – thanks to William Murdoch – gas illumination.*
Science Photo Library

alter the nature of the gas and also on how to safely purify, transport and store the result. In 1798 he returned to Birmingham where he continued experimenting and illuminated part of the interior of the Soho Foundry building. In 1802 to celebrate the Peace of Amiens, Murdoch illuminated the exterior. Three years later he fully lit the Philips and Lee cotton mills in Salford, initially with 50 gas lights but with continuing experimentation – using lime to purify the gas to remove the smell, for example – this soon increased to 904. But, Murdoch made a fatal mistake: he did not patent his work and never really benefited from it. He remained an inventor and partner of Boulton & Watt until 1830.

33 COTTON GIN
ELI WHITNEY – GEORGIA, USA – 1793–94

The cotton gin speeded up the method of processing cotton, thus inadvertently instigating a social revolution that made a living hell of the lives of millions of African Americans in the Deep South.

Massachusetts native and Yale graduate Eli Whitney (1765–1825) needed to make money to pay his debts, so he moved south to work as a private tutor on a plantation in Georgia. He quickly saw that to make upland short cotton profitable, cotton planters needed an easy way to separate the sticky green cotton seeds from their surrounding fluffy cotton balls. Supported both financially and morally by his employer Catherine Greene (widow of Revolutionary War general Nathanael Greene), he decided to invent a suitable machine knowing that he could patent the result for 14 years (it is currently 20 years) and become rich on the proceeds. Over a period of a few months, he devised the cotton gin (gin being short for engine).

The device was a simple mechanism that pulled the raw cotton balls through a set of wire teeth mounted on a revolving wooden cylinder and hooks that pulled the fibre through rows of narrow comb-like apertures that blocked the seeds passing through. It could clean up to 25kg of cotton per day. His first version was hand-cranked, and later bigger models were horse- or water-powered. This so vastly speeded up the process that his patent (applied 1794, validated 1807) was widely flouted and Whitney made virtually no money from the idea despite taking the infringers to court numerous times. He went out of business in 1797.

The economic and social ramifications of the cotton gin were enormous: by the mid-19th century the USA grew and supplied three-quarters of the world's cotton and this made Southern planters in particular very rich. In turn this increased

◀ *Portrait of Eli Whitney. Although his cotton gin proved hugely successful, Whitney never made any money from it as his patent was widely flouted. He spent years pursuing infringements, but there were too many to prevent and the profits from cotton too considerable to stop Southern planters exploiting him.* Library of Congress

THE FIRST COTTON-GIN.—DRAWN BY WILLIAM L. SHEPPARD.—[SEE PAGE 814.

their rapacious desire for more land and increased their need for slaves. The numbers of slaves escalated: in 1790 there were six slave states, by 1860 there were 15. The 1860 US Census gives a total slave population of 3,953,761 – an estimated one in three Southerners were slaves.

▲ *An inadvertent effect of the cotton gin was the callous exploitation of compelled labour: slavery. Millions of captured Africans were transported across the Atlantic to work on the pitiless cotton plantations in ever-greater numbers as cotton was planted over vast swathes of the southern United States. Even after Britain renounced slavery, its links to the South ensured continuing complicity.* Library of Congress

◄ *Cotton-picking needed cotton pickers – even more so after the cotton gin was invented. Coloured lithograph after Barfoot.* Wellcome Foundation (CC BY 4.0)

34 HYDRAULIC PRESS

JOSEPH BRAMAH – BARNSLEY, ENGLAND – 1795

The hydraulic press enabled engineers to undertake designs and manufacturing methods that would otherwise have been exceedingly difficult – if not impossible – to perform.

▼ *A painting of English inventor and locksmith Joseph Bramah (1748–1814) by an unknown artist. He was born in Stainborough, Barnsley, Yorkshire.* Science Photo Library

One of the problems that the engineers of the Industrial Revolution had to overcome when they started building things on a larger scale was that much of the tooling they had just wasn't powerful enough to cope with the demands placed on it. A fundamental process in the construction of many components requires that one part is pressed onto another – the friction fit they form then keeps them from coming apart. When that something is made twice the size of a previous version, however, it requires a lot more than twice the pressure to make it. The result is a problem that can be overcome only through access to more powerful presses.

Inventor Joseph Bramah (1748–1814) from Barnsley, in Yorkshire, England, came up with the solution when he developed the hydraulic press. A manufacturer of high-quality locks, he was used to designing his own machine tools. A very large part of his success was down to the fact that he insisted on a very high standard of workmanship and component inspection. This resulted in the production of accurate and reliable machinery, and he is today seen as the founding father of industrial quality control.

Bramah's press exploited Pascal's principle of fluid dynamics: that the pressure in a closed system remains constant throughout. To exert the necessary force, he moved the piston in a small cylinder back and forth to slowly pressurise a larger one – on the end of which was the ram that did the actual pressing. In this way the slow application of force was magnified such that what seemed like minimum effort could impose an enormous load.

Bramah patented his idea in 1795, and it was such a success that we know the machine as the Bramah Press to this day. Versions of it can be found in almost every workshop across the world.

▶ *The early Industrial Revolution was an era when many basic engineering problems held back progress. One of these was that the size and complexity of mechanical designs was often limited because manufacturers lacked the tools to be able to press things together at anything more than rudimentary pressure. Joseph Bramah's hydraulic press solved this problem overnight.* Wellcome Collection

PRESS.

Mʳ BRAMAH'S HYDROSTATIC PRESS.

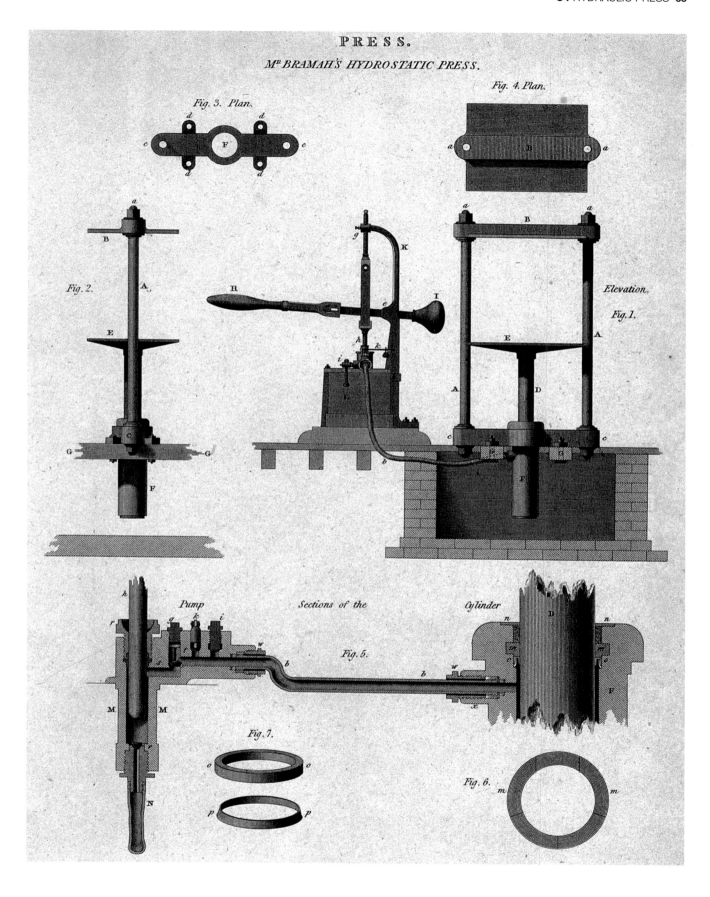

Fig. 3. Plan.

Fig. 4. Plan.

Fig. 2.

Elevation.

Fig. 1.

Pump

Sections of the

Cylinder

Fig. 5.

Fig. 7.

Fig. 6.

35 CAST-IRON AQUEDUCT
THOMAS TELFORD – SHROPSHIRE, ENGLAND – 1796

In seeking to replace a stone aqueduct destroyed by flooding, Thomas Telford constructed the world's first significant iron-built aqueduct.

▼ Designed by Thomas Telford, the aqueduct at Longdon-on-Tern represented the first such structure to be constructed in cast iron. Although the canal is no longer operational, the aqueduct survives and is a Grade I listed building. Alamy

Although considered to be the world's first cast-iron aqueduct, the 57m-long structure on the Shrewsbury Canal at Longdon-on-Tern in Shropshire was in fact completed about a month later than the much shorter (13m) span built by Benjamin Outram (1764–1805) at Holmes on the Derby Canal.

An Act of Parliament in 1793 authorised the construction of a canal east from Shrewsbury towards Ketley, Trench and Newport. The 27km Shrewsbury & Newport Canal (or just Shrewsbury Canal), with its 11 locks and inclined plane at Trench (built to provide a physical link with the Wombridge Canal), was working by 1797, but was isolated from the national network until 1835 when the Birmingham & Liverpool Junction Canal completed a link from Norbury Junction to Wappenshall.

The Shrewsbury Canal's initial chief engineer was Josiah Clowes (1735–95), but he died during its construction and was replaced by Thomas Telford (1757–1834), who was a Shropshire County Surveyor, and was also involved in the design of the nearby Ellesmere Canal.

In taking over the Shrewsbury Canal project, Telford was immediately faced with a problem. The stone-built aqueduct that Clowes had constructed to cross the River Tern at Longdon-on-Tern had been washed away in floods during February 1795. Initially Telford considered a replacement stone structure. However, the influence of the local ironmasters, such as William Reynolds (1758–1803) – many of whom had invested in the canal company – persuaded him to use cast iron instead. The resulting structure has two cast-iron troughs some 57m long. Cast in sections at Reynolds' ironworks at Ketley, the larger of the two troughs – 2.3m wide and 1.4m deep – was designed to hold the canal while the narrower side trough was to act as the towpath. The new aqueduct was completed in 1796.

The Shrewsbury Canal became part of the Shropshire Union Railways & Canal Co. in 1846, but the rise of the railway industry gradually led to its decline (along with many other canals) and the first closures along its watercourse occurred in 1922. By 1939 the canal west of Longdon had closed and the final section of the route was closed in 1944. Although some sections of the canal have been lost altogether, the aqueduct at Longdon-on-Tern – now Grade I listed – remains extant and, in 2000, the Shrewsbury & Newport Canals Trust was established with the aim of preserving and restoring the route.

36 SCREW-CUTTING LATHE
HENRY MAUDSLAY – LONDON, ENGLAND – 1797

Although screw-cutting lathes had existed since the 1500s, none was accurate enough for commercial manufacture. Maudslay's design was the first to allow full-scale production of bolts that were consistent enough to be interchangeable.

The basic principle of the lathe – that is, a machine that can turn an object on an axis while a stationary tool cuts into it – has been known since ancient times. Toolmakers have made various advances over the years, with early screw-cutting versions – where a single point tool is guided along the workpiece in a controlled

◀ *A portrait of engineer, toolmaker and inventor Henry Maudslay, by Pierre Louis Grevedon, dated 1827. Maudslay was born in Woolwich, London and initially trained as a blacksmith, following in his father's footsteps to work in the Royal Arsenal.* Science Photo Library

▼ *Maudslay's lathe provided for the first time a machine tool that was robust and consistent enough to manufacture interchangeable parts. The model seen here is a later Richard Roberts version.* Getty Images

manner to cut a thread – existing from around the 1400s. These implements were relatively primitive, though, and not capable of the accuracy required for industrial use. As several cuts are required to make a thread of a useable depth, the biggest technical problem was finding a way to ensure that the tool always travelled along the same path when subsequent cuts were taken. If its positioning was slightly out, the tool cut the wrong portion of the material and ruined the thread.

The answer was to have an accurate leadscrew – essentially a threaded bar that ran parallel to the lathe's bed. The cutting tool would be linked to this via an indexing locator, and as long as it was always engaged in the right place, the cutting tool would repeat the previous cutting path. Leonardo da Vinci drew designs along these lines – some even had change-wheels allowing the gearing to be altered so that the machine could cut different thread pitches. Nothing from this era ever actually resulted in a practical machine tool though.

In the 1700s various engineers put forward new designs, but none was suitable for industrial applications. Henry Maudslay (1771–1831), however, invented a lathe in 1797 that combined all the required features in a package that was practical and robust enough for everyday usage. Made on a wooden carcass, it opened the doors for the production of components that were accurate enough to be interchangeable. Consequently, it wasn't long before the introduction of recognised standards so that everyone could make items with the same threads. The standards listed all the salient aspects, such as the thread pitches, angles, major and minor diameters, and so on.

From then on lathes became popular in production workshops where they were subject to intense stresses day after day. Maudslay realised that the machines needed to be stronger in order to maintain accuracy and function. He followed up his screw-cutting design with the first all-metal lathe in 1810.

37 DITHERINGTON FLAX MILL
SHROPSHIRE, ENGLAND – 1797

In providing a solution to the problem of mill fires, the building of the flax mills at Ditherington in Shrewsbury resulted in the construction of the world's first iron-framed building – the forerunner of today's skyscrapers.

Now considered to be the grandfather of the skyscraper, the flax mill constructed at Ditherington, close to the centre of Shrewsbury, can rightfully claim to be one of the most important industrial buildings ever constructed in the British Isles. Historically, one of the major issues with earlier mills was their tendency to burn down. Industries such as the cotton trade generated vast quantities of dust. This, combined with candle lighting and the presence of other naked flames, made for an explosive combination and fire was an ever-present danger.

Towards the end of the 18th century, a number of architects and civil engineers were attempting to find a solution to this problem. William Strutt (1756–1830) had used cast iron in the construction of bridges and had introduced cast-iron columns into the construction of mills in Derby and at Belper where he designed the new West Mill completed in 1795. Here, the cast-iron columns supported wooden beams, cased in sheet iron as a protection against fire, with brick-built arches supporting the floors.

Strutt's work was taken a stage further in the construction of the Ditherington flax mills. He retained the cast-iron columns and brick arches, but at Ditherington he replaced the wooden crossbeams with beams constructed in iron, giving us the world's first iron-framed building. The building was designed by Charles Woolley Bage (1751–1822), a native of Derbyshire who had come to live and work in Shrewsbury, on behalf of the Leeds-based mill owner John Marshall (1765–1845) and his partners, the brothers Thomas and Benjamin Benyon, both Shrewsbury wool merchants. The trio had formed a partnership to construct a flax mill in Leeds in 1795, but the destruction of this by fire the following year was the impetus they needed to construct a new fireproof mill at Ditherington.

The mill was built during 1796 and 1797 and was completed for the sum of £17,000. The partnership between Marshall, the Benyon brothers and Bage ceased in 1804 with John Marshall becoming the sole owner of the mill. It continued to process flax until 1886, and after its sale spent the bulk of the next 100 years in use as a maltings for the brewing industry before this work finally ceased in 1987. Derelict for a number of years and placed on English Heritage's 'Buildings at Risk' register, the site – which is now all either Grade II or Grade II* listed – was acquired by English Heritage and the complex is now under a long-term restoration scheme.

► Internally the flax mills at Ditherington had a cast-iron framework allied to brick inverts. The primary aim of the bricks was to minimise the risk of fire: many earlier mills had been destroyed when the dust resulting from the manufacturing process combusted. Peter Waller

► External view of the Ditherington Flax Mills complex, near Shrewsbury. The building, through its pioneering use of cast iron in its construction, is now considered to be the forerunner of the modern skyscraper.

38

PUFFING DEVIL
RICHARD TREVITHICK – CORNWALL, ENGLAND – 1801

Richard Trevithick was the son of a Cornish mine manager who became one of the greatest engineers and inventors of the Industrial Revolution and the first man to build a working steam locomotive.

With his heritage, Richard Trevithick (1771–1833) was always going to work on the mines. However, his interest in and gift for engineering and innovation led him to focus on how Cornish tin mines could make use of steam engines. He was continually modifying and improving the new technology, particularly in boiler design and construction, always aiming to make them safe enough to cope with higher steam pressures. His experimentation led him to devise high-pressure steam engines that eliminated the need for a condenser and so could be smaller, lighter and more compact. Also, without a condenser, the engines needed less water, a crucial factor where supplies were limited.

At home in Camborne, Trevithick made working models of his high-pressure steam engines, initially static engines and then engines attached to a road carriage. A vertical pipe (chimney) vented the exhaust steam (no need for a condenser) and a crank converted the linear motion into a circular one. He had to be careful how he worked so as not to infringe Boulton & Watt's patented steam engines (see page 52) as these rival engineers were continually on the lookout for patent violations. (This was a new technology where a lot of money could be made and the competition was ruthless.)

In 1801 Trevithick unveiled the *Puffing Devil*, a full-sized steam road carriage that he demonstrated locally driving up Camborne Hill on Christmas Eve. The engine broke down three days later during further tests, unable to sustain enough steam pressure. It was left nearby while everyone went to the local public house for roast goose and drinks, but the water boiled off, everything overheated and the entire carriage burnt out. Trevithick was apparently not unduly concerned: he continued experimenting and refining steam locomotion for the rest of his life. A statue of him holding a small-scale model engine stands proudly on a plinth in Camborne outside the public library, sited on part of the route he took with the *Puffing Devil*. The last Saturday of every April is 'Trevithick Day' in Camborne, a celebration of his discovery of high-pressure steam power in particular, and of Cornish mining and industrial heritage in general.

▶ *Dubbed the 'Cornish Giant', Richard Trevithick was one of the leading lights of the early years of the Industrial Revolution. His leap of invention and imagination saw him devise and make the first locomotive steam engine using high-pressure steam to provide the power. Although a prolific inventor, this immense man – 1.9m tall and a renowned wrestler – died in poverty.*
Getty Images

▲ *A demonstration model of Trevithick's high-pressure steam engine. The engine itself could produce steam pressure of around 345kP. It was more compact and powerful than previous steam engines and made locomotive power a reality.* Getty Images

◄ *Detail from the plinth of the Trevithick Memorial in Camborne.* Tim Green from Bradford/ WikiCommons (CC BY 2.0)

39 CAPE HATTERAS LIGHT
NORTH CAROLINA, USA – 1802

One of the most dangerous stretches of eastern seaboard coastline was made considerably safer with the construction of the Cape Hatteras Light.

▼ *The location of Cape Hatteras on the eastern seaboard of North Carolina is shown on the bottom left of the map.* New York Public Library Digital Collection

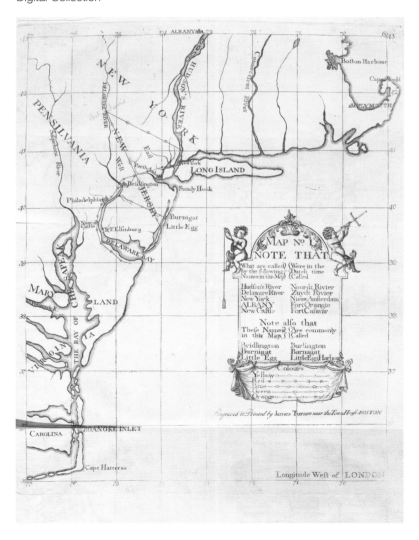

On Hatteras Island in the Outer Banks barrier islands at the far west of the Atlantic Ocean sits Cape Hatteras Light. Here, the warm, northward-flowing Gulf Stream meets the cold, southward-flowing Virginia Drift, itself a branch of the Labrador Current. The currents force southbound shipping into an area of treacherous, shifting, 19km-long sandbars called the Diamond Shoals. This is an area of powerful storms and swells where thousands of ships have been wrecked over the centuries – not for nothing was it called the 'Graveyard of the Atlantic'. The area was so notorious that Congress authorised the building of the lighthouse in 1794 and allocated $44,000 for the project. It is said that Founding Father and first Secretary of the Treasury, Alexander Hamilton, knew the area as a child and was so frightened as he once sailed around the cape in stormy seas and pitch-black conditions that he vowed that, if he ever had the opportunity, he would make sure a lighthouse was built on Cape Point to save lives at sea.

Construction began in 1799 and ended in 1802 on ground 34m above sea level. Originally of dark sandstone, the lighthouse rose to 27.5m high, but proved hard to see in anything but clear conditions, being too short and too dark. It had 18 whale-oil lamps with 36cm reflectors. In clear weather the light was visible for 29km. The first lighthouse keeper was Adam Gaskins, appointed by President Jefferson himself, a year before the project was finished.

By the 1850s sea captains regularly complained that Hatteras was one of the worst lighthouses on the Atlantic coast. They said it was too short, and that the light was too dim and as much a hazard to shipping as a help, as it was easily mistaken for the light of another vessel. So, in 1853 the tower was extended by 18.5m to become 46m tall, and the extension was painted red to stand out. Finally, the lighthouse was given a new Fresnel lens (see page 54) to throw the light further out to sea.

In summer 1861 Federal troops occupied Hatteras Island with a number of soldiers billeted around the lighthouse to protect it from a Confederate plan to blow it up. Shelling damaged the lighthouse itself, worsening its already deteriorating structure. Congress decided to replace Hatteras lighthouse in 1867.

▼ *In 1999 the controversial decision was made to move Cape Hatteras Light away from the encroaching Atlantic Ocean that was only some 15–20m distant. Despite some local protests, the lighthouse was jacked up, then rolled on rails to its new location 450m back from the ocean. Since then its former foundations have on occasion been covered by seawater.* Library of Congress

40 PAPERMAKING MACHINE
FROGMORE MILL, HERTFORDSHIRE, ENGLAND – 1803

The ability to manufacture paper on a large scale meant that books, previously the province of the rich, suddenly became affordable. This resulted in a huge increase in literacy and made education much more widely available.

▼ *Two illustrations of the paper-making process. Producing paper the old way (*below*) by hand was slow and expensive. Using the methods of mass production (*bottom*) reduced labour costs.*
Science Photo Library

Paper is so commonplace these days that few give it a thought. However, until modern production methods, its manufacture was a slow and expensive process. Although there were several alternatives – papyrus and parchment being the predominant ones – all were extremely costly and in short supply. As a result it was only religious institutions and very wealthy individuals who could afford books. All this changed when the first papermaking machine was invented at Frogmore Mill, originally a corn mill, in the village of Apsley, Hertfordshire.

The story behind the installation of the machine is a complicated one. It was originally invented by a French accountant called Nicholas Louis Robert (1761–1828). However, he sold the designs to his employer, who in turn passed them on to an Englishman, John Gamble. Gamble then took out an English patent and teamed up with Henry (1766–1854) and Sealy Fourdrinier (1773–1847), who had the financial means to make it all happen. They commissioned an engineer, John Hall, to build the actual machine. Later joined by his brother-in-law, Bryan Donkin (1768–1855), the papermaking machine was completed in 1803.

The process started with the use of a dilute pulp suspension that was made up of shredded linen and water. This was then poured onto a revolving wire mesh barrel that drained off some of the water as the pulp was fed into a press. There, it was transferred onto a felt blanket that was then fed between two rollers that squeezed out any remaining water. This left it dry enough to be rolled up before being cut into sheets and hung out to dry. The system was revolutionary, and the reduction in labour costs meant that the price of paper fell by 75%.

Later improvements to the design saw the addition of things like steam driers, making the process even quicker. By the end of the 19th century, Great Britain was manufacturing 650,000 tons of paper a year and saw a huge increase in literacy.

ORUKTOR AMPHIBOLOS

41

OLIVER EVANS – PHILADELPHIA, USA – 1804

Although the *Oruktor Amphibolos* was not a commercial success, conceptually it was a massive step forward, showing that locomotives and steamboats were possible, and thus inspiring a new generation of inventors.

The *Oruktor Amphibolos*, as its inventor Oliver Evans (1755–1819) called it, was both the first steam vehicle and the first amphibious vehicle in the USA. Although steam engines had been in the country for some time, they were mostly low-pressure versions that were both low-powered and very inefficient.

◄ *An engraved portrait of Oliver Evans by W.G. Jackman. Known as the 'Watt of America', when he died in 1819, Jackman's empire stretched from Pittsburgh to Philadelphia as his ironworks produced steam engines for use all over the USA.* Science Photo Library

▼ *This illustration from* The Boston mechanic and journal of the useful arts and sciences *published in July 1834 shows the* Oruktor Amphibolos: *the boat/vehicle that Evans invented. Although it was not a commercial success, its use of engines powered by high-pressure steam showed that locomotives and steamboats were possible.* Getty Images

Evans was a firm believer that high-pressure steam was the way forward, and had long-held dreams of being able to produce steam-powered vehicles. The problem was that using high pressures required a major technical leap forwards in terms of the manufacturing process and materials. On the positive side, such engines were not only a lot more powerful, but far smaller than conventional low-pressure versions.

Evans was not deterred by the seemingly insurmountable technical issues involved with using high pressures – especially the belief that boilers could not withstand them – and he forged ahead with new designs. He quickly realised that the conventional method of using condensers to control the steam was inefficient, and replaced them with a variety of new mechanisms including double-acting cylinders and new valve arrangements. The radical changes resulted in engines that were simpler to build, as well as being both cheaper and easier to run. A further benefit was that they operated on a lot less water, which meant that they were far more suited to installation in vehicles, such as automobiles and locomotives, and much more practical for a wide variety of industrial uses.

Evans' first commercial venture with the engine was for the Philadelphia Board of Health who needed a dredger to clear the city's dockyards and to remove sandbars. The design he came up with was the *Oruktor Amphibolos* (amphibious digger), a flat-bottomed vessel that used bucket chains to undertake the dredging part of the operation. Powered by one of Evan's high-pressure engines, it was nearly 9m long, 3.7m wide and weighed in at 17 tons. In order to get the boat from his Philadelphia workshops to the Schuylkill River, he added four wheels, powered from the same source. On 13 July 1805, it made the journey – and the record books. Unfortunately, it appears that it wasn't good at its job, as it was broken up for parts three years later.

42 MEER ALLUM MULTIPLE ARCH DAM

HENRY RUSSELL – HYDERABAD, INDIA – 1804

The first multiple arch dam in the world, Hyderabad's Meer Allum Dam was probably the first dam of any kind to depend for its stability on the arch principle.

▼ *The artificial lake was years in the planning but took only two years to build. It opened in June 1806 and construction was so sound that it did not need wholesale repair until 1980. It supplied much of Hyderabad's drinking water for 125 years before being replaced by the much bigger Himayat Sagar and Osman Sagar reservoirs. Dome.mit.edu*

Hyderabad, in central southern India, has a monsoon season that arrives from the southwest in the middle of June and lasts until about 1 October. It dumps around 81cm of rain across the region in just those months (the remainder of the year is bone dry). The precious water collects in 'tanks' (reservoirs) that then provide a year-round supply of water for both people and crops. Meer Allum (also known as Mir Alam) Dam was built to supply and regulate such water for the city.

Details of the dam's gestation are vague. In the late 18th century Britain and France were fighting for dominance in India. It seems that

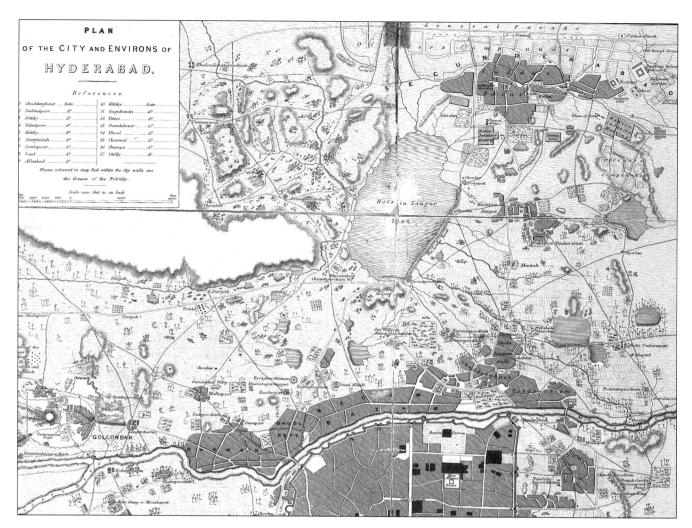

PLAN
OF THE CITY AND ENVIRONS OF
HYDERABAD.

References.

the French started work on the dam and then the British took over and finished it. The dam was built across the Musi River near Hyderabad with money given by the Nizam of Hyderabad to the prime minister, Mir Alam, as a reward for winning the fourth Mysore War against Tipu (the Sultan of Mysore who supported the French).

The original plan for the dam was (according to the Hyderabad Water Works) designed by the French engineer Michel Joachim Marie Raymond (1755–1798), but he died years before building actually started. When the British took over, Henry Russell (1783–1852), assistant to the Resident of Hyderabad, took on the role of dam supervisor and he is generally credited with its ingenuity. Mir Alam Bahadur laid the foundation stone on 20 July 1804 and construction was completed by 1808. At the time it was estimated the tank could contain 10 million m^3 of drinking water for Hyderabad.

Constructed by the Madras Engineering Corps, the dam was built to a unique design that uses abatement reaction forces to stay in place – the arches provide the strength and stability of the structure. The result is a concrete curved dam 915m long that incorporates 21 semicircular individual vertical arches with vertical upstream faces, all of the same thickness, but with varying spans ranging from 42m to 24m. A major part of the thrust-forces acting on the dam are transmitted to the arches. The overflow is partially discharged through a spillway at one end, while the rest pours over the crest. Remarkably, there has been no damage to either the mortar or masonry and the foundations have remained solid.

No one copied the design of the dam until 120 years later when the Coolidge Dam in Arizona was built using the same principles.

▲ *The water in the Meer Allum tank was said to be so sweet that locals carried it with them whenever they left Hyderabad. The tank became a well-known tourist attraction (and remains so), with visitors advised to collect and carry its water with them on their journeys.* Dome. mit.edu

43 OYSTERMOUTH RAILWAY
THE MUMBLES, WALES – 1804

There had been industrial and goods railways for some time before the Oystermouth Railway became the first railway in the world to carry paying passengers.

Although railways had existed for more than a hundred years at the start of the 19th century, the lines that had been built were designed exclusively for the movement of goods. All that was to change in 1807 when the Oystermouth Railway introduced the world's first services for fare-paying passengers. The line had originally been authorised by Act of Parliament as the Oystermouth Railway & Tramroad Co. in June 1804. The original Act stated that the line could be operated by 'men, horses or otherwise'. The 4ft gauge line – built primarily for the movement of coal from collieries in the Clyne Valley and limestone from the Mumbles to the Swansea Canal – opened in 1806.

One of the line's shareholders, Benjamin French, proposed the introduction of a passenger service and this first operated – using one or converted wagons – on 25 March 1807. To the west of Swansea on Swansea Bay, the Mumbles was a potentially popular destination for day-trippers from Swansea and French was prepared to pay £20 per annum for the rights to operate a horse-drawn passenger service. However, the construction of a

► *Although passenger services over the Oystermouth Railway commenced in 1807, they were suspended in 1826 as a result of competition from coaches using a newly opened turnpike road. This view records the horse-operated service reintroduced in 1877, following the conversion of the line to standard gauge. Subsequently both steam and electric services were operated.* Barry Cross Collection/Online Transport Archive

competing turnpike road mean that by 1826 the passenger service was suspended. This was not, however, to be the end of the story.

In 1840 John Armine Morris (1812–93), one of the sons of the first chairman – Sir John Morris (1775–1855) – sold the line, which was effectively moribund, to his brother George Byng Morris (1816–99) and, 15 years later, the line was rebuilt to standard gauge. Five years later horse-operated passenger services were reintroduced. In 1877 – faced by competition from the Swansea Improvements & Tramways Co., which had introduced competing horse trams to the line in 1874 – the Oystermouth began using steam traction. In 1879 the line was renamed the Swansea & Mumbles Railway Co. Ltd and a further change of traction saw electric tramcars replace the steam-hauled services in March 1929. Utilising the largest first-generation tramcars built for operation in the British Isles, the Swansea & Mumbles celebrated its 150th anniversary in 1954, but six years later the line was closed, to be replaced by the diesel buses of South Wales Transport.

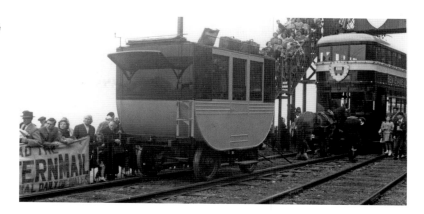

▲ *In 1954, in order to mark the 150th anniversary of the authorisation of the Oystermouth Railway – and the 147th anniversary of the first passenger services – a replica of one of the early horse trams was built. It is seen in the company of one of the massive electric trams that operated over the line until it closed in 1960. The replica tram is now preserved and on display in Swansea.* Barry Cross Collection/Online Transport Archive

▼ *This Railway Clearing House map illustrates well the complex nature of the railway network serving Swansea in the early 20th century. Many of the lines – including the Oystermouth (which, by this date, had become the Swansea & Mumbles) – had been built to exploit the mineral wealth of the region and move coal and other raw materials to the booming docks in Swansea.* via Peter Waller

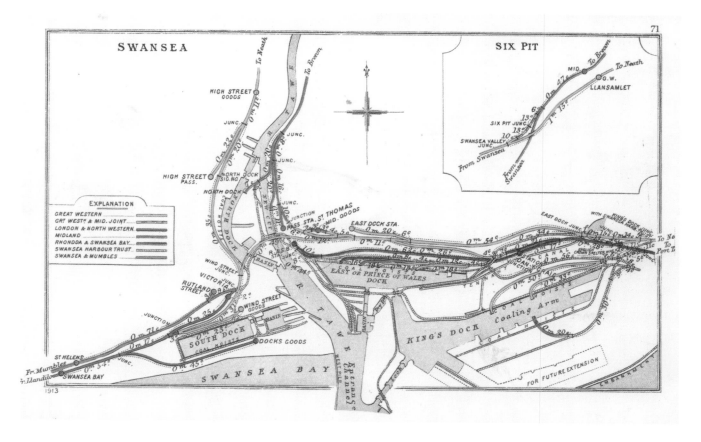

44 DUNDAS AQUEDUCT
JOHN RENNIE – SOMERSET, ENGLAND – 1805

It may not be the first aqueduct in Britain, but it's certainly one of the most handsome and a glorious example of the way that the Industrial Revolution combined modern ideas with an artistic eye for detail.

▼ *Owing to continuous problems with leakage the canal was closed in 1954 and abandoned, drying out altogether in the 1960s and 1970s. But after campaigning and fund raising, the canal was restored and reopened in 1984. Today, it is a wonderful long-distance footpath and bridleway through the heart of England. The etching is by John Shury after William Williams.* Wellcome Collection

This handsome Georgian aqueduct carries the Kennet & Avon Canal across the Avon Valley and over the River Avon near Monkton Combe, Somerset, where it forms a junction with the Somerset Coal Canal. It was named after Charles Dundas (1751–1832), the MP for Berkshire, who advocated that a junction connecting the Kennet and Avon rivers would make great financial sense. He was also instrumental in raising the funds to pay for it. In honour of this, the aqueduct was named after him and he also became the first chairman of the Kennet & Avon Canal Company.

In 1788 plans were made for a western extension for the Kennet Canal and engineer

The Dundas Aqueduct, Claverton
NEAR BATH.

◄ *The Dundas Aqueduct carries the Kennet and Avon canal over the River Avon and in 1951 became the first canal structure to be designated a Scheduled Ancient Monument.* Robert Powell/WikiCommons (CC BY-SA 3.0)

John Rennie (1761–1821), who surveyed the main line of the canal and also the line of the Somerset Coal Canal, was appointed to oversee the Dundas project. Construction work started in 1797 under his supervision and that of chief engineer John Thomas (1752–1827). Work was completed by 1805. The specific purpose of the aqueduct was so that coal-carrying barges, coming west from the Paulton colliery in north Somerset, were able to transport their cargo directly into the industrial heart of England via the network of canals that crossed the country.

Built of local ashlar Bath stone, the aqueduct is 137m long and comprises three rusticated arches. It is embellished with giant Doric pilasters and balustrades on both sides at either end. The flanking oval side arches each span 6m, while the much larger central, semi-circular arch spans 20m. The top parapet is pierced with balusters over revetment walls. It has a triglyph frieze cornice broken forward over pilasters.

The water in the aqueduct was pumped by the Claverton Pumping Station and water mill at nearby Limpney Stoke. The latter contained a wheel 7m wide and 5m in diameter, driven by water from the River Avon that flows at 2 tons of water per second and which in turn powered the beam engine pump, also designed by John Rennie. This lifts the water over 15m up to the canal at a rate of 450,000 litres/hr.

After years of service the aqueduct was closed in 1954 because of extensive leaks and damage. It was abandoned and used solely as a walkway. However, by the 1980s sufficient interest in the revival of canals led to the restoration of the aqueduct and it reopened again in 1984 after being relined. The water is now provided by electric pumps.

▼ The aqueduct crosses the railway and is well used by walkers, cyclists and canal boats. Arpingstone/ WikiCommons

45 PERCUSSION IGNITION
ALEXANDER JOHN FORSYTH – SCOTLAND – 1807

Until 1807, all firearms suffered from a significant time delay between pulling the trigger and the gun going off – leading to inaccuracy. When Forsyth introduced percussion ignition, the problem disappeared.

Alexander John Forsyth (1769–1843), a Scottish Presbyterian clergyman, was a keen hunter. Frustrated by the inefficiencies of his flintlock shotgun, he decided to find a way to improve it. This consisted primarily of

reducing the 'lock time' – the delay between pulling the trigger and the firearm actually discharging. He reasoned that the longer the lock time, the more inaccurate the shot – and the less food there would be for the table. His first successful design followed much experimentation and used what is known as the 'scent bottle lock'. This consisted of a small vessel that contained mercury fulminate, a chemical that explodes when struck. It replaced the pan and frizzen – parts that were normally needed on a flintlock – and it could hold around 20 shots-worth of fulminate.

The resulting gun was so successful that Forsyth was persuaded by Lord Moira, the Master of Ordnance at the Tower of London, to spend a year there developing it further such that it could be used by the military. Unfortunately, when Lord Moira was replaced, the new master unceremoniously told Forsyth to leave. It is said that Napoleon Bonaparte offered Forsyth £20,000 to take his invention to France, but he declined the chance to become a rich man overnight and stayed in the UK. Despite his setback with the British military, Forsyth was granted a patent on 11 April 1807 – this was cleverly worded and prevented anyone else from using his percussion design.

Forsyth then set up a successful business in Piccadilly in London – Alexander Forsyth & Co. – making firearms and converting old-style flintlocks to percussion ignition. Dissatisfied with his existing designs, in 1813 he further refined them, replacing the scent bottle with a sliding tube that deposited the requisite amount of fulminate next to the touch hole through which the discharge would be initiated when struck by the hammer.

Forsyth's designs were a milestone in firearms history, and from that moment on flintlocks were obsolete. All today's firearms use percussion caps that stem directly from his design.

▶ *The critical parts of a firearm where percussion ignition is concerned. The hammer **H** is pulled backwards until it is locked in place by the sear (not seen here). When the trigger is pulled, the hammer snaps forward striking the primer, **P**, causing a flame **Z** to issue forth into the main powder charge – located inside the chamber, **S**, and so firing a shot. Until this advance of 1807, all firearms suffered from a significant time delay between pulling the trigger and the gun going off, leading to inaccuracy both on the battlefield and on the hunting ground.* Getty Images

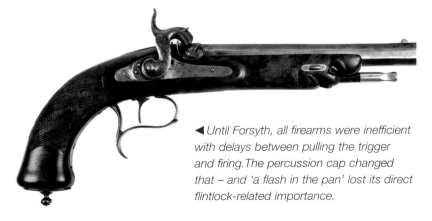

◀ *Until Forsyth, all firearms were inefficient with delays between pulling the trigger and firing. The percussion cap changed that – and 'a flash in the pan' lost its direct flintlock-related importance.*

46 GAS STREET LIGHTING
FREDERICK WINSOR – LONDON, ENGLAND – 1807

Until street lighting was installed it was dangerous to wander through a city like London. Lighting ensured that theatres, restaurants and other places of public entertainment thrived as people felt safer going out and about after dark.

Frederick Albert Winsor (1763–1830) was a German inventor interested in gas lighting whose dream was to light the streets of London. In 1803 he demonstrated coal gas lighting at a lecture at the Lyceum Theatre on the Strand. That year he patented the invention

▼ *'A Peep at the Gas Lights in Pall-Mall' by Thomas Rowlandson in 1809. The background shows Carlton House, the home of the Prince Regent. The cartoon shows the sensation that gas lighting caused among Londoners. The lady on the far right (a prostitute) is worried that such illumination will harm her business, while the man on the left is explaining to his companion how the gas works. There were many sceptics about the technology, including Sir Humphrey Davy, who said 'it would be as easy to bring down a bit of the moon to light London as to succeed in doing it with gas'.* WikiCommons

and moved to live in two houses on fashionable Pall Mall, central London, where he carried out numerous experiments with gas.

In 1806 Winsor presented a paper to the Royal Society explaining his ideas for gas lighting. The following year, knowing that he needed publicity to catch King George III's attention, he placed gas lights between Pall Mall and St James's Park along the walls of Carlton Palace Gardens for the king's birthday celebrations. Using gas generated by the furnaces inside his house on the Mall, he projected images on the walls of the houses. A little later he erected 13 lampposts along one side of Pall Mall from St James's to Cockspur Street, again using gas from his home pumped through a wooden pipe buried under the pavement. Pall Mall became the first street in the world to be illuminated by gas, which was noted by observers to have 'much superior brilliancy'. However, the gas illuminated only a few feet around the post and there were still areas of darkness between the lamps.

Parliament refused Winsor permission to start a national gas company. He repeated the demonstration several more times in the following years as he tried to persuade MPs of its efficacy. Winsor and his supporters had to wait until 1812 for parliament finally to charter his Gas Light and Coke Company (the first public gas company) to supply for 21 years the cities of London and Westminster and the Borough of Southwark, using coal gas light from the gasworks in Horseferry Road.

On 31 December 1813, Westminster Bridge was lit by gas. Street lighting proved such a success that many other companies fought for permission to lay gas pipes and light the streets and within 15 years almost every large town in the country was lit. Baltimore became the first city in the USA to get gas lighting in 1816, followed by Paris in France in 1820. By 1823 40,000 gas lamps illuminated 346km of London's streets.

47 PORTLAND OBSERVATORY
LEMUEL MOODY – MAINE, USA – 1807

One of the earliest observation towers in the USA, the Portland Observatory was engineered and constructed to a unique design. It remains the only historic maritime signal station in the United States.

At the turn of the 19th century, Portland in Maine was a busy port, valuable for its deep harbour. However, ships entering the port were hidden from sight until they rounded the point of land at Spring Point Ledge, by which time they were almost at the docks. Former sailor Captain Lemuel Moody (1768–1846) had the inspiration to construct a tall tower on Munjoy Hill, 68m above sea level, to serve as a communication and observation post. He planned to use a powerful telescope to spot incoming ships and for an annual subscription of $5 would alert ship owners to give them time to prepare for their cargo's arrival and make special accommodations as necessary.

Moody made his proposal for the observatory and eight subscribers came forward. On 20 March 1807 they signed an agreement for the building of a 'Marine Lookout' on 'some elevated part of Mountjoy's Neck'. It was not to cost more than $2,000. They began a subscription to issue 100 shares each at $20 (to a maximum of seven shares per person) to raise the funds: 55 proprietors bought the 100 shares.

Construction started in 1807, with shipwrights making a significant contribution to building the tower, as the joinery and structure testify. Looking somewhat like a lighthouse, the tower rises up seven storeys to 26m high and is built with a tapering octagonal design to lessen the impact of the prevailing winds. It has a 9.8m-diameter base that tapers to 4.6m at the observation

◀ *Photograph of the Portland Observatory in 1936, some 13 years after it ceased its working life. Today it is one of the biggest tourist attractions in Maine, welcoming hundreds of visitors between May and October every year. It remains the last standing marine signal tower in the USA.* Library of Congress

▲ *The view from the Portland Observatory is a great panorama of the harbour and it shows how the low-lying outer islands and headlands obscured the view of incoming ships from ground level.* Brian Feathers/WikiCommons (CC BY-SA 3.0)

deck. Each octagonal corner has a solid post of local Maine white pine like a ship's mast. The ground floor is a grid of substantial timbers filled with 122 tons of loose rock ballast (collected from nearby fields). The rock stabilises the tower during storms and high winds. The cupola (lantern) at the top held a P. & J. Dollond achromatic refracting telescope that could identify ships up to 50km away.

To inform his subscribers of their ship's arrival, Moody used a system of signal flags hoisted up flagstaffs. The flags were also used to communicate back and forth with the incoming ships. Moody manned the Observatory himself from 1807 until 1846.

The tower greatly increased the efficiency of Portland harbour and it remained a working marine signal tower, run by the Moody family, until 1923 when the invention of the two-way radio made it obsolete. During the War of 1812, the Observatory was used as a watch tower. After this, it fell into disrepair.

▶ *Elevation of the Portland Observatory. The elaborate wooden structure was mostly made by shipwrights who used locally sourced wood.* Library of Congress

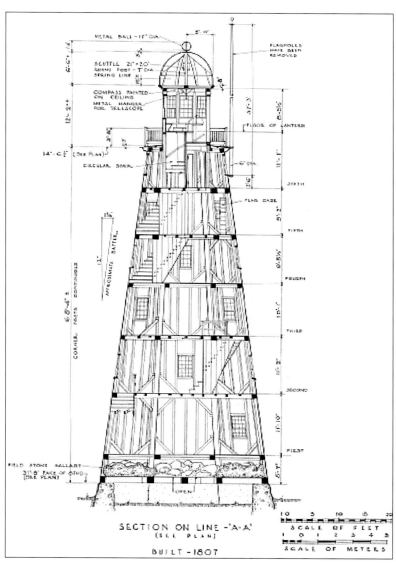

SECTION ON LINE "A-A"
(SEE PLAN)

BUILT - 1807

48 | BELL ROCK LIGHTHOUSE
JOHN RENNIE – FIRTH OF FORTH, SCOTLAND – 1807–11

The Bell Rock Lighthouse lies around 18km east of Dundee in the North Sea, between the Firths of Forth and Tay. It warns of a long and treacherous reef that has claimed many vessels and thousands of lives.

▼ *The great artist J.M.W. Turner immortalised the Bell Rock Lighthouse and showed a fascinated public an insight into the extreme wind and wave conditions the lighthouse had to withstand. He also shows a ship under full sail warned away from the rock by the tower's faithful light.* Library of Congress

Also known as the Inchcape Rock, the red sandstone reef runs roughly 610m across the shipping route into the Firth of Forth and the approaches to Dundee and the River Tay. It is covered with water, averaging 3.7m deep at high tide, and exposed only at low water – making this a very difficult build, hugely dependent on weather and tide conditions.

Robert Stevenson (1772–1850) was surveyor to the Commissioners of Northern Lights and drew the initial plans. John Rennie (1761–1821) was appointed chief engineer and assisted Stevenson – exactly

▶ *Low tide on a calm day reveals part of the reef on which the lighthouse – also known as the Inchcape Light – stands. It is now the oldest surviving sea-washed lighthouse in the world.* Kognos/WikiCommons (CC BY-SA 4.0)

how much has been the subject of much debate (particularly by their two sons). Work started in 1807 and an estimated 110 men and one horse (Bassey) built the lighthouse.

It is built from sandstone quarried from Mylnfield, near Dundee and Craigleith, and Edinburgh granite from Rubislaw, Aberdeen and Cairngall (near Peterhead). It rises to 35.3m high, and has a diameter of 12.8m at the base and 4.6m at the top. In all it uses some 2,835 stones. It was constructed using solid dovetailed masonry for the first 10m – half of this is below high water – with courses 1 to 26 having an outer casing of granite and a sandstone inner. Courses 27 to 90 are sandstone only; the outer grouting was of Roman cement. Above, lie six chambers: from the base – the provision store, the lightroom store, accommodation for three light keepers, the kitchen and eating room, the strangers' room, and the library. The top of the lightroom is made of cast iron with copper and brass fitments.

The first optical system used 24 parabolic reflectors with inner silvered surfaces, each 63.5cm in diameter. These were arranged in a rectangle with seven on each of the two widest sides, arranged in three rows 2:3:2. The other ten reflectors were fitted to the minor sides and had red glass discs around the outer rim arranged 2:1:2. At the focus of each reflector was an Argand lamp (see page 54) fuelled with spermaceti oil, with a circular wick almost 2.5cm in diameter. Clockwork revolved the whole array, itself activated by a heavy weight descending through the tower. It transmitted a unique red-and-white signal light identifying the Bell Rock Lighthouse for up to 56km. The first revolving lighthouse in Scotland, it alternated between a red and white light every four minutes and an entire rotation took eight minutes. In the 1820s the reflectors were replaced with the latest 1st Order Fresnel lens (see page 54) and paraffin vapour burners.

The total cost for the original lighthouse was £61,331 9s 2d exactly.

49 CAEN HILL LOCKS
JOHN RENNIE – WILTSHIRE, ENGLAND – 1810

These locks were the ingenious solution to convoy barges up and down a steep incline.

▼ *Over this 3km stretch, the locks are, of necessity, close together, especially in the middle of the rise. This also means that the side pounds are unusually capacious to be able to store sufficient water to operate the locks.* BazViv/WikiCommons (CC BY 3.0)

The most problematic section of the 140km-long Kennet & Avon Canal is the steep rise of Caen Hill near Devizes. Consequently, it was the last section to be built. The solution, worked out by John Rennie (1761–1821), the canal's chief engineer, was a magnificent flight of 16 hill locks. They form the second longest continuous flight of locks in the country, but the steepest rise over the shortest

distance – a gradient of 1 in 30 – and can take up to six hours for a vessel to clear.

The locks convey canal barges up the side of the hill between Reading and Bath along a 3km section of the canal that contains a total of 29 locks that shepherd the canal waters up a rise of 72m. Logistically, to get the boats up this section, the locks had to be really close together, but this posed the problem that they would quickly run dry owing to an insufficient volume of water. Also, emptying the locks risked causing flooding further downstream. To solve these two dilemmas, Rennie dug long, rectangular retention pounds (reservoirs) of 2,833m² each, between the locks on the north side of the canal. These stand by, ready to feed water into the lock system as required and to collect water as the locks empty. The pounds were lined with up to almost 1.2m-thick puddle clay.

The bricks for the lock chambers were dug from clay deposits to the south of the canal.

In the hours of darkness between 1829 and 1843, for an extra charge of one shilling per barge and sixpence per boat, the flight of locks was lit by gas lights.

Finished in 1810, the last section finally completed the Kennet & Avon Canal between the important ports and business centres of London and Bristol. However, the canal provided a vital link for only about 40 years, until it was superseded by the opening of the Great Western Railway in 1851.

The canal had cost over £16,000 per mile to build, but it never proved as profitable as the investors hoped. At its peak in 1815, it carried around 151,980 tons of cargo annually, but made its first loss in 1877 and never went into profit again. The last load was carried from London to Bristol in 1900. The canal fell into disuse and was closed. However, after decades of dereliction and neglect, significant restoration work repaired the locks and they reopened in 1990.

▲ *John Rennie surveyed the 140km course of the Kennet & Avon Canal, routing the last section through the towns of Newbury, Trowbridge and Devizes. Between Devizes and Rowde, the steep hill required an extraordinary 16 locks to take the waters in a straight flight up Caen Hill.* Rwendland/ WikiCommons (CC BY-SA 4.0)

50 TIN CANS

BRYAN DONKIN – LONDON, ENGLAND – 1811

The development of tin cans for preserving food allowed travellers, sailors, explorers and others to eat healthily while away from home. The world's first can-making factory opened in Bermondsey, London, in 1813.

During the Napoleonic Wars, it became ever more important to find a way to feed soldiers and sailors properly while they were far away from home. The French government offered a 12,000-franc prize for anyone who could solve the problem. French chef Nicolas Appert (1749–1841) won the money in 1810. In *L'Art de Conserver*, he demonstrated how to safely preserve meat and vegetables using airtight cork-sealed glass bottles, then sterilising them in boiling water. He used the money to set up a bottling plant to preserve food for French soldiers. However, the system wasn't perfect: the glass bottles were heavy and easily broken.

Another Frenchman, Philippe de Girard (1775–1845), began experimenting with iron canisters for food, but failing to find support in revolutionary France, he contacted a London-based agent, Peter Durand, who registered the patent in England. Durand used this as a basis to investigate using iron cans covered with a layer of tin to prevent corrosion and was issued with a patent of his own in 1810. But the process was lengthy: food was packed in the cans, then sealed, put into cold water and gradually brought up to boil for hours, then the lid was reopened slightly and resealed. Each can was stored at a temperature of 32–43°C before leaving the factory.

Taking this further, in 1811 English mechanical engineer Bryan Donkin bought Durand's patent for £1,000 and started experimenting with various canning techniques of his own. By 1813 he was in partnership

◄ *Bryan Donkin and his brother-in-law opened their first canning factory in 1813, supplying canned food to the Royal Navy and British Army. But it was the publicity gained by supplying canned goods to Arctic explorers that really caught the attention of the British public.*
Getty Images

▲ *A canning kitchen. Donkin's factory industrialised the process. Early cans were much larger than today's versions and required a hammer and chisel to open. A consignment of cans from the William Parry expedition was abandoned in the Arctic, but eight years later, in 1833, saved the lives of the members of the struggling John Ross expedition.*
Science Photo Library

with John Hall (his brother-in-law) and John Gamble and together they formed a food-canning manufacturing company (Donkin, Hall & Gamble) in Bermondsey, south London. Their first commercial cans were produced in summer 1813. Cans could be anything from 2kg to 10kg in weight and needed a hammer and chisel to open. The average cost was around 4s 9d/kg, too expensive for ordinary people to afford.

Their enterprise impressed the Duke of Kent (later George IV) and with his endorsement the process of canning really took off. The British explorers who set off to search for the Northwest Passage took their cans of preserved meat with them; and William Parry took them on all four of his 1820s voyages to the Arctic.

In 1813 the Admiralty bought 70kg of canned food and initially used them to nourish sick sailors. Soon they were ordering more – beef, mutton, soup, parsnips and carrots – and by 1821 they were ordering more than 4,000kg of tinned goods a year.

Donkin left the company to pursue other innovations, primarily papermaking machinery, after which Hall & Gamble rapidly expanded their range of canned goods.

51 THE NATIONAL ROAD
USA – 1811 ONWARDS

The National Road – the first Federally funded highway, built between 1811 and 1837 – began to tame the vast distances of the USA helping to knit together the far-flung reaches of the fledgling country.

▼ *A freestone culvert, some 32km west of Jonesville, Virginia, illustrated by J.K. Wharton. The National Road provided jobs for labourers and craftsmen not just for road building, but also for the numerous businesses, particularly inns and taverns and supply stores that sprang up and thrived along its course.* New York Public Library Digital Collection

In the early years of its existence, the size of the new country and the lack of quick communications made it difficult to promote unity and commonality of purpose, especially for the new isolated settlements in the West. Both Thomas Jefferson (1743–1826) and George Washington (1732–99) agreed that they needed a trans-Appalachian road, running between Cumberland, Maryland, and the Ohio River. Eventually, on 29 March 1806 Congress authorised the project and President Jefferson signed the Act to establish the National Road (at the time called the Cumberland Road).

The contract was signed in 1811 for the first 16km of roadway that headed westwards and

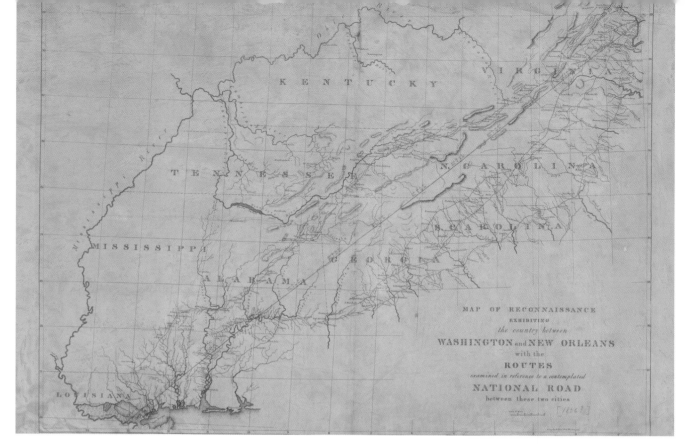

MAP OF RECONNAISSANCE
EXHIBITING
the country between
WASHINGTON and NEW ORLEANS
with the
ROUTES
examined in reference to a contemplated
NATIONAL ROAD
between these two cities

was completed in 1813. Seven years later, the road had reached Wheeling, West Virginia, where the building stopped. It was some 1,319km long and mail coaches started to run the route. To the east, by 1824, the road had pushed to Baltimore.

Some years later road building resumed, taking a route through central Ohio and Indiana, then on to reach Vandalia, Illinois, in the 1830s. During this period the National Road became the first US highway to use the new macadam road surfacing. At the same time the government passed over responsibility for much of the road to the relevant states, who built toll houses and toll gates to collect revenue – although the Federal government retained responsibility for road repairs. The financial crisis of 1837 halted plans to push the road to St Louis and construction stopped again.

Small settlements, stores, livery stables, inns and taverns (an average of two per 3km) sprang up along the length of the route, while the destination towns like Uniontown and Washington, Pennsylvania, and Wheeling, West Virginia, became business and commercial centres. The road itself gathered popular names such as the National Pike and particularly Main Street, a name that lodged into popular culture.

▲ *Map of the reconnaissance exhibiting the country between Washington and New Orleans with the routes examined in reference to a contemplated national road between these two cities. The road system – also often referred to as Main Street – was built between 1811 and 1834 and was the first Federally funded road in the USA. It featured in popular songs and entertainments and was responsible for starting the mass movement of people westwards.* New York Public Library Digital Collection

The National Road flourished through the 1820s, stuttered through the economic difficulties of the late 1830s, and revived in the 1840s when travellers moved westwards in their thousands, journeying in stagecoaches (averaging 100km a day) and covered wagons, to settle in the Ohio River Valley.

Brightly painted Conestoga wagons rattled along the road travelling about 25km a day, transporting the likes of sugar, coffee and various staple goods to settlers in the East, and returning loaded with produce from the frontier farms.

The heyday lasted until the arrival of the railroads in the 1870s.

▶ *Milepost on the National Road between Tridelphia and West Alexander showing distances east to Cumberland and west to Wheeling.*

52

STANLEY MILL
JOSEPH WATHEN – GLOUCESTERSHIRE, ENGLAND – 1812–14

Although not the first metal-framed mill in England, Stanley Mill is the finest, rising five storeys above the Cotswold countryside – an area celebrated for its textile production.

▼ *The metal fabric of the mill links the practicalities of ironwork with the Victorians' unerring eye for aesthetics to produce a building that is important historically and visually.* www.whateversleft.co.uk

The Cotswolds are known for their sheep. The wool industry made the fortunes of many, giving rise to the proliferation of 'wool churches' in the area. King's Stanley, a village on the outskirts of the Gloucestershire town of Stroud, benefited from the arrival of Flemish weavers and clothmakers in the 14th century and, by the 18th, a significant number of workers with hand looms were employed locally.

In around 1811–12, Joseph Wathen started building a new mill on the site of an older one (there's some evidence that the site dates back to the 12th century). Iron-framed to help withstand fires (the elegant columns and trusses were made by Benjamin Gibbons of Dudley), the fireproof construction was well underway by the time Wathen sold the site to George Harris and Donald Maclean in 1813 for the princely sum of £8,655.

The new owners completed a number of extensions and additional workshops, producing an impressive range of buildings. The main L-shaped block is built of stone and brick around the iron frame, with cast-iron pillars on each floor supporting the brick vaults above. The windows have multi-

◄ *The mill was originally water-powered with five water wheels. However, the vagaries of the water source meant that the mill required a more predictable power source, so a Boulton & Watt engine was installed.* www.whateversleft.co.uk

paned iron casements; the cast-iron columns are notable for their quality and the spandrels take power-drive shafts. Heavy metal doors contribute to the fireproofing and the building withstood a major blaze in 1884.

Power for the mill originally came from five water wheels fed by a 2-hectare mill pond. A 40hp Boulton & Watt steam engine (see page 52) was added in 1824. In 1834 the five water wheels, with a fall of 5m, generated power equal to 200hp. At the time the mill is said to have employed between 800 and 900 people.

The Admiralty used the mill during World War II, but it was finally closed in December 1989. It survives in good order and there is some talk of its development.

53 *PUFFING BILLY*

WILLIAM HEDLEY – DURHAM, ENGLAND – 1813–14

Puffing Billy was a vitally important link in the development of the steam engine as an effective means of propulsion. It established the principle that locomotives could haul a significant weight by adhesion only.

In 1810, during a strike that affected the Durham coalfield, the then owner of Wylam Colliery, Christopher Blackett (1751–1829) took the opportunity to experiment to see if a wagonway could be operated by a locomotive simply using adhesion. Although there had been attempts to use a steam engine earlier, these had proved inadequate. Nonetheless, during the

► *Designed by William Hedley,* Puffing Billy *was restored to its original condition in 1830 following the reconstruction of the railway at Wylam Colliery. It is now preserved as a static exhibit in the Science Museum in London.* Getty Images

◀ *A side view of* Puffing Billy *in its original condition. Three similar locomotives were built to Hedley's 1813 design, two of which were used on the railway that served Wylam Colliery in Northumberland. The second locomotive (*Wylam Dilly*) incorporated slight improvements on* Puffing Billy *and is preserved in the National Museum of Scotland in Edinburgh.* Getty Images

Napoleonic Wars, a shortage of horses gave renewed impetus to the work. On the Middleton Railway, John Blenkinsop (1783–1831) patented in 1811 a rack-and-pinion system whereby a geared driving wheel engaged with teeth cast into the cast-iron rail to improve adhesion.

Blackett was keen to see if traction via simple adhesion was possible. Initially he tested a hand-cranked wagon fitted with a central driveshaft and geared drive. This was successful and the first engine – possibly based on the locomotive design by Richard Trevithick (1771–1833) for the Pen-y-darren Railway – was constructed. This was not sufficiently powerful to work efficiently, but it established that there was potential for adhesion-operation steam locomotives.

The result was three engines, the first of which was *Puffing Billy*, constructed between 1812 and 1814 by the engineer William Hedley (1779–1843), who was the resident engineer at Wylam Colliery, the engine-wright Jonathan Forster (1775–1860) and the blacksmith Timothy Hackworth (1786–1850).

The locomotive incorporated a number of features that were patented by Hedley. It had two vertical cylinders, one on each side of the boiler, which drove a crankshaft that coupled the driving wheels together (for the first time) to provide better traction. The new locomotives were, however, too heavy for the track. As a result in 1815 they were rebuilt with four axles to give a better weight distribution. They were again rebuilt back to their original two-axle form after 1830 when improved track meant that their original weight was no longer a problem.

The locomotive remained in use at Wylam until it was presented on loan to the Patent Office Museum in London by the then owner of Wylam Colliery. This museum – the forerunner of London's Science Museum – was later to purchase the locomotive and it remains on display at South Kensington. It is now the oldest surviving steam locomotive in the world.

Another of Hedley's locomotives built for the Wylam Railway (*Wylam Dilly*) is also preserved and now on display at the National Museum of Scotland in Edinburgh. This locomotive had an interesting subsequent career: in 1822 it was mounted on a ship's keel and used to provide power to a small paddle steamer that transported strike breakers. After that, it returned to colliery work at Wylam.

The influence of Hedley's work was significant. Among those to benefit was George Stephenson (1781–1848), who lived locally and drew upon this knowledge when developing his own pioneering locomotives.

54 DAVY LAMP

HUMPHREY DAVY – YORKSHIRE, ENGLAND – 1815

Mining is a dangerous industry made more dangerous by flammable gases. The safety lamp allowed miners to work in considerably less perilous conditions underground and saved many lives.

Mining was a dark, dirty and dangerous business, but it was one of the biggest industries in Britain, particularly since the harnessing of steam to drive the machinery of the Industrial Revolution. Conducted in the bowels of the earth, mining was one of the most physically risky jobs of its time. The arrival of the Davy Lamp made the process no less arduous, but at least it made it considerably safer.

In 1815 Humphrey Davy (1778–1820) introduced his safety lamp, designed for use in flammable atmospheres where firedamp or minedamp (mostly methane) lurked, such as in coal and tin mines. It was used for the first time at Hebburn Colliery, then in County Durham, in 1816. Davy was not the first to devise a safety lamp; William Clanny (1813) and George Stephenson (1815) had demonstrated examples, but Davy's was the most effective.

▲ The Davy Lamp warned miners of dangerous levels of lethal gases in a mine through the blue flaring of the flame. But the lamp also inadvertently contributed to high levels of mortality among miners because it made them more complacent and liable to explore previously out-of-bounds areas. Getty Images

▶ Portrait of Sir Humphrey Davy in the Wellcome Trust Museum. He was knighted in 1812 as one of Britain's leading scientists and was invited to Paris (despite the war) to receive a medal from Napoleon. He was President of the Royal Society between 1820 and 1827. Wellcome Collection

Davy's lamp surrounded the lamp flame with a mesh screen that arrested the flame – the holes in the mesh were large enough for air, but too fine to allow a flame to pass through and ignite any firedamp in the atmosphere. The lamp could also indicate the presence of gases: if flammable gas was around, the flame inside the lamp flared up with a blue tinge. The lamp incorporated a metal gauge to measure the height of the flame, so miners could assess the level of danger. When lowered near the ground it could detect pockets of carbon dioxide and other gases that are denser than air. When the carbon dioxide was near to dangerous levels (known as black damp or choke damp), the lamp would extinguish itself. This would happen before the levels of gas became lethal, so miners had time to retreat from danger of asphyxiation. The lamps were not infallible, however: a damaged gauze would affect the reliability of the security system, while low airflows and inadequate ventilation in the mine would mean that the flame could fail to register the gases.

Davy formally presented a paper describing his lamp to the Royal Society in London that November. For his discovery he was awarded the Society's Rumford Medal and £2,000 in silver raised by public subscription.

The invention did not entirely prevent mining explosions, of course. In fact, in 1835 a Select Committee on Accidents in Mines reported that the lamp had led to more deaths because it encouraged the exploration of dangerous passages that had previously been left alone for safety reasons. Additionally, as miners supplied their own lights and as the Davy Lamp emitted only a feeble light, many miners chose to use a candle flame anyway. Mine regulations often forbade candles, but this was impossible to enforce. Still other miners liked the use of both the candle and the Davy Lamp.

55 FIRST MACADAMISED ROAD

JOHN MCADAM – BRISTOL, ENGLAND – 1816

The single biggest advance in road building since the Romans, McAdam's innovative road-surfacing material improved the longevity of the road, provided a smoother surface and simplified the process of road construction.

John Loudon McAdam (1756–1836) was born into an aristocratic family in Ayr, Scotland, and became interested in roads (in particular the theory and practice of road construction) while a young man. His first successful road was a stretch on his own estate between Alloway and Maybole. In 1787, he became a trustee for the Ayrshire Turnpike in the Scottish Lowlands during which time his interest in roads became an obsession. After living in Bristol for some years, he was elected Bristol's Surveyor-General in January 1816, becoming responsible for 240km of road for the Turnpike Trust. At last he could put his theories to the test with the first 'macadamised' stretch of road, Marsh Road at Ashton Gate, Bristol.

Soon he was a Trustee for 34 different road trusts and immersed in road construction techniques. He approached parliament with improvement ideas for roads, and three times provided evidence for parliamentary road enquiries. He wrote two influential treatises: *Remarks on the Present System of Road-Making (with observations, deduced from practice and experience, with a view to a revision of the existing laws, and the introduction of improvement in the method of*

◄ *Portrait of John McAdam, who did much not only to improve road surfaces, but also to expose the corruption of turnpike toll trusts.*
Hulton Archive/Getty Images

making, repairing, and preserving roads and defending the roads from misapplication) and *Practical Essay on the Scientific Repair and Preservation of Roads*, 1819.

McAdam instructed that his roads should be levelled directly into the soil but above the surrounding ground and above the water table. The width was to be 9m with a rise from the edges to a 7.6cm camber, so that water could run off to ditches on either side. The lowest 20.3cm layer was broken aggregate of stones no larger than 7.6cm. This was topped with a 5cm layer comprising stones under 1.9cm in size. No permeable materials were allowed. Sitting labourers using small hammers smashed the stones into pieces no heavier than 170g, while supervisors carried scales to measure and check the stones. The stone size was important so that those on the surface were smaller than normal carriage-wheel width. The stones then had to be carefully and evenly spread across the road surface, a shovel-full at a time, giving a solid layer of broken angular stones. No surface binding was allowed – passing traffic would compress the surface.

McAdam's road construction method was quick and economical and proved immediately and immensely popular. His ideas spread rapidly and road travel became much faster, smoother and more widespread.

McAdam's insistence on intelligent management of the roads and their regular maintenance was also enormously influential. He wanted a central road authority with a salaried professional official who would be immune to bribes and responsible and answerable for the roads. Tarmac, named in tribute to McAdam, was patented by Edgar Purnell Hooley (1860–1942), a Welsh civil engineer and inventor, in 1902.

56 MILLING MACHINE
ELI WHITNEY – NEW HAVEN, USA – c1818

The milling machine is such an integral part of most manufacturing processes that without its invention the Industrial Revolution as we know it could not have happened.

Up until the advent of the milling machine, the removal of metal from an object was either done by hand with a saw or a file, or on a lathe. 'Rotary filing', as it was termed, basically involved putting a rotary cutter into a lathe and then moving the object – known as the 'workpiece' – slowly past it. The cutter would then remove any metal it came into contact with.

The significance of the milling machine is that it can undertake machining operations that are

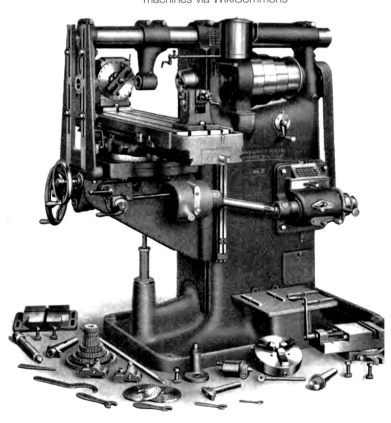

▼ *The horizontal milling machine was invented around 1818 and differs little conceptually from the advanced systems in use today. The machining operations that suddenly became possible when it became available radically altered the way most industrial items were made, reducing cost and increasing production. This is a machine from the end of the 19th century.* From Practical treatise on milling and milling machines via WikiCommons

either difficult or impossible to achieve either by hand or on a lathe. In its simplest form, it holds a rotating cutter in either the horizontal or vertical position. The object to be machined is then clamped to a sliding table and slowly moved past the cutter. Depending on the materials involved, the depth of cut required and so on, the workpiece may be moved back and forth many times, each pass taking another cut. In its early forms milling was considered a roughing operation to reduce the amount of hand filing, rather than replacing it entirely. Later refinements, however, would make the mill accurate enough to dispense with any direct manual involvement at all.

Once in widespread use, milling machines became so important to most industrial processes that it became unthinkable to set up any kind of manufacturing business without access to one. Although they are far more complex these days and many have incredibly advanced computerised control systems, the essential aspects have not changed since they were first invented.

The credit for their invention usually goes to Eli Whitney (1765–1825), who introduced his machine around 1818, although many others were working on similar designs at the same time. Best known for inventing the cotton gin (see page 62), he was a talented inventor who also worked on firearm design. It is no coincidence that the main development centres of early milling machines were within the armaments industry, especially at the two Federal armouries at Springfield and Harpers Ferry, as well as at private establishments.

▼ *The Eli Whitney Armory was one location where work led to the creation of the milling machine. For many years credited to Whitney himself, the invention was more likely to have taken place at Springfield, Massachusetts, or Harpers Ferry, West Virginia.* Library of Congress

57 HETTON COLLIERY RAILWAY
COUNTY DURHAM, ENGLAND – 1822

Through its use of steam engines, stationary engines and inclined planes, the Hetton Colliery Railway in County Durham was the first in the world that did not require the use of any animal power.

By the start of the 19th century, the concept of constructing wagonways for the movement of coal from colliery to the river staithes (wharves) in Northumberland and Durham for onward shipment was well established. Individual landowners, however, sought to develop mines on their own land as knowledge of the great northeastern coalfield grew. The Hetton Coal Company was established in 1819 to extract coal from land owned by Thomas Lyon and his son John, to the south of Houghton-le-Spring. In order to transport the coal, they decided to construct the 13km-long Hetton Colliery Railway. George Stephenson (1781–1848) – noted for his work on other wagonways – designed the railway. Using a combination of stationary engines, self-acting inclines and steam locomotives, the line was the first ever to be worked without any animal power and was Stephenson's first entirely new line.

The line's resident engineer was George Stephenson's brother, Robert (1788–1837). He oversaw construction, laying the first new track in March 1821. The line officially opened on 18 November 1822. In order to obviate the need for animal power, two stationary engines raised wagons over Warden Law Hill, while five self-acting inclines, where the weight of descending wagons helped raise the empties, were also completed. In addition, George Stephenson supplied five steam locomotives that were 0-4-0 with chain-coupled wheels, and were modified from original designs with springs. These could compensate for the action of the vertical cylinders that had caused locomotives to rock and damage the cast-iron track. At the line's terminus on the River Wear, staithes were completed in order to load the coal directly into ships.

The original line was not wholly successful and Robert Stephenson was sacked in 1823. Engineers then made a number of improvements, including installing a third stationary engine at Warden Law Hill. When the line finally closed on 12 September 1959, it was regarded as the oldest mineral railway in Britain.

▶ *Reproduced in an 1826 publication,* The American Farmer *by William Strickland, this image shows both a typical Hetton Colliery train along with a lineal explanation of the route that the line's engineer – Robert Stephenson – built.* via Peter Waller

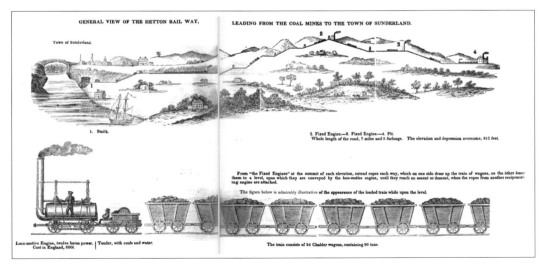

58 | *AARON MANBY*
CHARLES NAPIER – BIRMINGHAM, ENGLAND – 1822

Aaron Manby, named for the engineer who conceived her, was the first iron steamship to go to sea and the first to provide direct passage from London to Paris using steam power – a herald of the future of shipping.

Captain, later Admiral Charles Napier (1786–1860) was a visionary Scottish naval officer who dreamed of building iron warships. As a precursor to this, he had the ambition of running a fleet of steamships up and down the River Seine. Ready and able to finance the investment, he recruited pioneering Staffordshire engineer Aaron Manby (1776–1850) and his son Charles (1804–84) to help him. Together they set about designing a prefabricated iron steamship. The pieces were made at Manby's Horseley Ironworks and then shipped in pieces to a works at Rotherhithe on the River Thames to be assembled. This was the first steamboat to be built on the 'knock-down' principle.

The PS *Aaron Manby* had a flat-bottomed hull 36.6m long made of 6.35mm-thick iron plate fastened to angle-iron ribs. She had one wooden deck, a bowsprit and a distinguishing 14.3m-high funnel. She was powered by Aaron Manby's patented (in 1821) oscillating engine, a steam engine he had designed specifically for maritime use. This drove two 3.7m-diameter paddle wheels that were only 0.7m wide to keep her width within the 7m working limit of the Seine. She could make 9 knots and drew 0.3m less water than any of her steam contemporaries. She weighed 116 tons burthen. Sceptics expected her to sink.

A little over one month after sea trials were completed on the Thames on 30 April 1822, the *Aaron Manby* crossed the English Channel on 10 June. Helmed by Captain Charles Napier and with Charles Manby as ship's engineer, she arrived at Le Havre having made an average speed of 8 knots with a few passengers and cargo of linseed and iron casting. She then steamed up the River Seine to Paris where she caused a sensation. She repeated the Channel crossing a few times, then was used for pleasure trips up and down the Seine.

Napier financed five similar iron steamships but his business failed and he was declared bankrupt in 1827. The *Aaron Manby* was sold to the Compagnie des Bateaux à Vapeur en Fer, a French consortium who based her at Nantes and operated her on the River Loire until she was decommissioned and broken up in 1855.

The use of iron plates instead of wood for the hull of the *Aaron Manby* revolutionised shipbuilding. She was a direct ancestor of HMS *Warrior*, the Royal Navy's first iron frigate, built in 1860, the year Napier died.

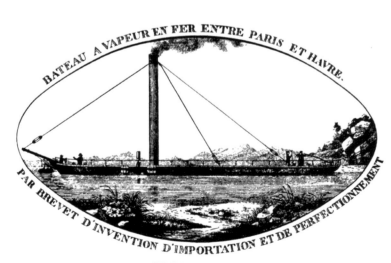

P.S. "Aron Manby," 1822

◀ *The* Aaron Manby *was the first iron steamship to cross the English Channel when she arrived at Le Havre in 1822.* WikiCommons

59

DIFFERENCE ENGINE
CHARLES BABBAGE – LONDON, ENGLAND – 1822

Although it had various precursors, the Difference Engine established the fundamental principles on which all modern computers run. It is therefore hard to overstate its significance in the annals of human history.

▶ *This image shows all that remains of Charles Babbage's original Difference Engine No.1, which is on display in the Science Museum, London.* Science Photo Library T404/0066

It is not known when the first mechanical computation device was built, but as a bronze navigation calculator was found in the remains of a 2nd-century-BC shipwreck it was certainly early in human history. More recent examples were proposed in the 17th and 18th centuries. However, they failed to become established.

The first real step towards modern computation since then came about when Charles Babbage (1791–1871) announced his invention of what he called the Difference Engine. This was presented on 14 June 1822 as a paper to the Royal Astronomical Society under the title *Note on the application of machinery*

▶ *An engraving of Charles Babbage published by* The Illustrated London News *on 4 November 1871, shortly after his death on 18 October.* WikiCommons

to the computation of astronomical and mathematical tables. It outlined an automatic mechanical calculator designed to tabulate polynomial functions, a task that was long, boring and so prone to error when undertaken by humans.

Such a machine could cope with a wide variety of complex mathematical functions including logarithms and trigonometric computations, making it extremely useful to engineers, scientists and navigators. However, his hand-cranked working model was a demonstration of only the basic principles – he needed to do a lot of research and development to build a full-scale version. He was granted funding from the British government – who were predominantly interested in getting better navigation tables – to build a full machine. Babbage soon discovered, though, that the technology of the era made it simply impossible to make the parts he needed with the requisite precision. He did manage to produce a small working model in 1832, but work on the bigger version was put on hold the following year. The government finally gave up hope and abandoned the project in 1842, by which time it had spent £17,000 on it.

Babbage went on to design a far more advanced version known as an Analytical Engine – this inherently made the original Difference Engine obsolete, although by 1849, he had also come up with the Difference Engine No.2, which could cope with 31-digit numbers and seventh-order differences. This was not only more powerful, but could also undertake the calculations more quickly while using fewer parts. Unfortunately for Babbage, his new and improved machine was outshone by a German version, and it faded from history.

60 ROBERTS LOOM
RICHARD ROBERTS – MANCHESTER, ENGLAND – 1822

The Roberts cast-iron power loom revolutionised textile manufacture for it provided the first reliable way to mass produce woven fabrics. Driven by belt from a steam engine, it featured many new mechanisms that made it easy to set up and run. Consequently, it soon became a mainstay of the British cotton industry.

▼ *This illustration shows the inside of a weaving area at a textile factory – the machines depicted are power looms with cast-iron frames, as developed by Richard Roberts. These were dimensionally stable even in the wet atmospheres required in such establishments (fibres snap if the air is too dry), minimising the risk of warping that had been rife with wooden frames, which made them economically inefficient.* Wellcome Collection

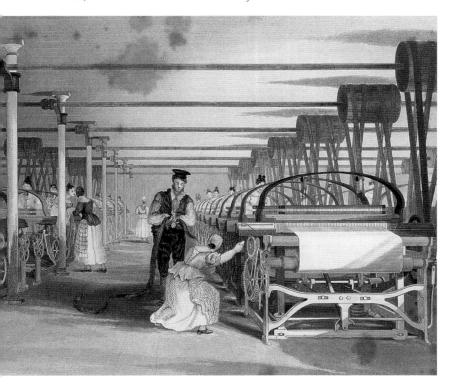

Although the first powered weaving loom was invented by Edmund Cartwright (1743–1823) in 1785 (see page 55), it was far from perfect. The main problem was that variations in the atmosphere's humidity caused the wooden frame to warp, creating all manner of problems with fabric consistency and in particular with yarn tensions. If such a thread becomes too slack, it risks getting caught in the machinery; if it is too tight, it will snap. Richard Roberts (1789–1864) overcame these carcase movement issues when he patented a cast-iron-framed power loom in 1822. He was an engineer who worked on producing precision machine tools for the textile industry, and as a result had the necessary expertise to design a robust solution.

Roberts' new machine not only had a cast-iron frame, but also included a number of other innovations. These were mostly intended to make it both easier to adjust and more reliable. One was a method of automatically taking up the changes in warp tension created by the fall in diameter of the warp beam (that is, the roll the thread comes off) as production progressed. Another involved a way of keeping the fabric tension correct as it came off the loom through a toothed wheel that worked a ratcheted pinion. The loom itself featured a main shaft that had a heavy flywheel to dampen out any vibrations in the series of drive belts, which were powered by steam. There were several other similar mechanisms that together made the machine extremely reliable and easy to run. As a direct consequence of its day-to-day usability, the Roberts Loom was widely taken up by the Lancashire cotton mills.

The large numbers of machines that were installed throughout the region revolutionised the manufacture of woven cotton. Once again, though, the resulting increase in weaving capacity brought about a second shortage of suitable loom threads.

61 PORTLAND CEMENT

JOSEPH ASPDIN – YORKSHIRE, ENGLAND – 1824

This fast-setting cement – the basic ingredient of concrete – was taken up wholesale by the building industry and was used extensively in new and imaginative building projects.

▼ *The blue plaque that celebrates the home in Angel Inn Yard, Leeds, of Joseph Aspdin, the inventor of Portland cement. His patent describes his mixture as 'My method of making a cement or artificial stone for stuccoing buildings, waterworks, cisterns, or any other purpose ... and which I call Portland cement.' Ben Dalton/WikiCommons (CC BY 2.0)*

Joseph Aspdin (1778–1855) was by trade a bricklayer and carpenter. Experimenting with materials in his kitchen, he discovered that by heating clay and limestone to a very high temperature, letting it cool, and then grinding the resulting aggregate and mixing it with water, he could make a particularly strong cement.

In October 1824 he was granted a patent for 'An Improvement in the Mode of Producing an Artificial Stone' – an innovation he liked to call Portland cement because it 'resembled the best Portland stone', an oolitic limestone from Dorset and the most prestigious building stone of the period. He used Pennine carboniferous limestone that was otherwise used for paving turnpike roads and town pavements. Initially Aspdin used offcuts (road scrapings) from stone used in road repairs. Indeed, he was twice prosecuted for removing whole paving blocks from the roads around Leeds.

Aspdin devised the mixture to be fast setting and low strength, ideal for architectural pre-cast mouldings and for stucco work for cornices and ceiling details. His cement was made by double burning a very pure limestone. First, he burned it on its own to make lime. Then he mixed it with clay and slaked, dried and crushed it, then burned it again in a vertical kiln until the carbonic acid was completely removed. The resulting calcine was then turned to powder. The exact proportions were a closely guarded secret.

Portland cement was manufactured in England and on the Continent after the patent was issued, but the process was time consuming and costlier than ordinary cement and the uptake was not great.

In 1825 Aspdin formed a partnership with a fellow Leeds man, William Beverley, to set up a production plant in Kirkgate, Wakefield. At the same time Aspdin obtained a second patent for a method of making lime. His youngest son, William, joined the company as an agent, but the pair fell out and William left in 1841.

William used a recipe with a higher limestone content, and burned it at a higher temperature (which used more fuel), then ground the previously discarded clinker and added it all to the mix. Two years later he set up his own cement plant at Northfleet in Kent, near plentiful soft chalk deposits and there developed a stronger, 'modern' Portland cement. William worked with Isambard Kingdom Brunel while Brunel and his father were working on the Thames Tunnel (see page 108). Brunel used this 'modern' Portland cement in what is thought to be the first large-scale application of the product. The tunnel was opened for the public to admire in 1843.

JOSEPH ASPDIN
(1778 ~ 1855)

Portland Cement, one of mankind's most important manufactured materials, was patented by Joseph Aspdin, a Leeds Bricklayer, on 21 October 1824. Aspdin lived in this yard (then called Slip Inn Yard) and first sold his cement in Angel Inn Yard.

62

THAMES TUNNEL
MARC BRUNEL – LONDON, ENGLAND – 1825–43

Marc Brunel's project to bore a tunnel beneath the River Thames demonstrated for the first time that it was possible to use a tunnelling shield successfully in the construction of tunnels beneath rivers.

▼ *Born in France, Marc Isambard Brunel is perhaps overshadowed by his more famous son Isambard Kingdom Brunel, but his achievements – most notably his development of the tunnelling shield and the completion of the Thames Tunnel – mark him out as possibly the more influential of the two in the long term.* Getty Images

The development of the canal network during the latter half of the 18th century had witnessed the construction of tunnels – such as the Blisworth Tunnel on the Grand Union Canal that was opened in 1805 after 12 years of hard work – but all of these involved work above water level. The genius of Marc Isambard Brunel (1769–1849) and his son, Isambard Kingdom Brunel (1806–59), was that they faced the threat of flood in order to pioneer the construction of a tunnel beneath an existing water course, and that tunnel – almost 175 years after its completion – is still playing a vital role in providing a link between the north and south banks of the River Thames.

There had been a long awareness that a crossing of the Thames downstream from London Bridge was essential. This could not be a bridge, as that would impede shipping to and from London's docks. Prior to Marc Brunel's involvement, there had been an earlier scheme, backed by Richard Trevithick (see page 70) but this had come to naught. Brunel had already thought about tunnelling – indeed he had proposed a scheme for St Petersburg – but it was in London where he saw his theories put into practice. He and the mercurial Thomas Cochrane, 10th Earl of Dundonald (1775–1860), patented the concept of a tunnelling shield in January 1818. Said to be based upon the shell of the *Toredo navalis* (or shipworm), the concept of the tunnelling shield is still used in all major tunnelling projects, such as the Channel Tunnel and the new tunnels serving Crossrail in London.

With the backing of many notable figures, including the Duke of Wellington, the Thames Tunnel Company was established in 1824 and work began the following year. The tunnelling shield – built by Henry Maudslay (1771–1831) at his Lambeth works (see page 67) – was eventually installed in November 1825 at Rotherhithe, following a series of mishaps. Boring began in

◀ *The Thames Tunnel – finally opened in March 1843 – marked the culmination of more than two decades of work. More than 170 years after its opening, it remains in use as a railway tunnel.*
Getty Images

November 1825. However, the tunnel workings were flooded twice – in May 1827, and again in January 1828 when six workers were killed – which stopped all progress. Work recommenced only in August 1835 with a new and improved shield replacing that installed in 1825. The work – impeded by four further floods, by fires and by problems of methane and other gases – was slow, but was completed in November 1840 and the tunnel finally opened to the public on 25 March 1843. The completed structure was 396m long, 11m wide and 6m high and ran at 23m below the river surface at high water.

Despite its technical triumph, the tunnel was a financial disaster. It had cost £634,000 to build – far in excess of the budget – and plans to make it available to vehicles as well as pedestrians were never progressed. In 1865 the tunnel was purchased by the East London Railway and converted to form a railway tunnel. The first trains operated through the tunnel on 7 December 1869 and, after many years as the East London line of London Underground, between 2007 and 2010 it was upgraded to form part of the new London Overground network.

63 ELECTROMAGNET

WILLIAM STURGEON – NEWFOUNDLAND, CANADA – 1825

The electromagnet is another seemingly simple device that has had huge repercussions on a wide variety of technologies – from the creation of telegraphic communications to the engines in today's washing machines.

▼ *The original drawing from William Sturgeon's 1824 paper to the British Royal Society of Arts, Manufactures and Commerce shows the key parts of his newly invented electromagnet. It was composed of 18 turns of bare copper wire through which passed an electric current. In the drawing, the eggcup-shaped bowls contained mercury – an early method of making an electrical contact; the left-hand arm worked as an on/off switch.* WikiCommons

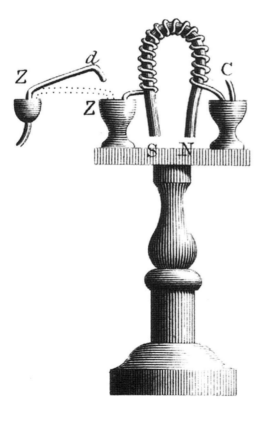

▼ *A contemporary portrait of William Sturgeon, English physicist and inventor. Best known for his work on the electromagnet, he also built and improved electric motors.* Science Photo Library

Danish physicist Hans Christian Oersted (1777–1851) first discovered electromagnetism in 1820, as part of an observation he made while lecturing on electricity. William Sturgeon (1783–1850), an electrical engineer who had been mesmerised by the thunderstorms he'd observed while serving in the army in Newfoundland, soon read about it. He became so fascinated by the phenomenon that he began a series of experiments and, in 1825, announced his invention of the electromagnet. This was a crude piece of iron in the shape of a horseshoe, around which he had wrapped several coils of wire. When he passed electricity through them, the iron became magnetised, and when he turned the current off, it demagnetised. He was able to show that an appropriately electro-magnetised piece of iron that weighed only 198g was able to lift a 4kg weight.

A further feature of Sturgeon's invention was that he could control the power of the electromagnet simply by varying the amount of electricity fed through it – the stronger the current, the greater the magnetic force it created. He went on experimenting and, in 1832, invented the commutator – this is the segmented device found on the end of most electric motors. Featuring in its simplest form two carbon brushes, it allows electricity to be fed into the rotating coils that drive the motor via electromagnetism.

American inventor Joseph Henry refined Sturgeon's work and made an even more powerful electromagnet. He was able to show that it could be controlled over a long distance, simply by applying an electric current – this was the basis of the telegraph, and thus it heralded the birth of modern communications.

Sturgeon went on to achieve great acclaim, becoming a well-respected lecturer and making many more significant inventions. Sadly, despite all his efforts and brilliant innovations, he did not make much money – he died in penury on 4 December 1850, in Manchester.

64 SELF-ACTING MULE

RICHARD ROBERTS – LONDON, ENGLAND – 1825

The advancements in weaving technology of the late 18th and early 19th centuries had created a insatiable demand for thread. Several people came up with supply solutions that allowed the spinning of cotton without constant skilled supervision. However, only one – Roberts' self-acting mule – proved practical in the commercial environment.

Not long after Richard Roberts (1789–1864) had invented the power loom for weaving (see page 106), he realised that unless more thread was spun, his market for the new machines would diminish dramatically. He therefore set himself to addressing the lack of spinning capacity in his usual, thorough manner by inventing a 'self-acting' spinning mule. It was what we would today call 'automatic', in that it needed very little supervision by the operator.

The mule he invented incorporated a number of novel features that automated it. These included a reversing mechanism on the top of each spindle, a falling guide that ensured the yarn went where it was wanted, and a control that turned the spindle at the correct speed as its diameter changed, along with various other yarn guide mechanisms. These were controlled by a series of levers and cams that were linked to a device known as the 'shaper'. The speed at which the spindles turned was modulated by a drum that was equipped with weighted ropes – this was turned by the mule's mainshaft and it engaged with the spindles via a toothed wheel. The whole system formed a package that, because of Roberts' robust engineering practices, proved not only very efficient and easy to run, but also very reliable.

Alongside his power loom, the self-acting mule hugely altered the employment scene of the time as they meant that neither the spinning nor the weaving required skilled craftsmen any more. Instead, semi-skilled labour could run the machines. This caused a major shift in the social environment as hordes of agricultural workers flocked to take up the new jobs that were on offer.

Typically, the spinning mules were operated by men, while the power looms were run by women. As the new self-acting equipment became the norm, the old hand-operated machines were increasingly referred to as 'mule-jennies'.

► *Spinning cotton with self-acting mules, which were belt-driven powered by water or steam engine. Note the child under the machine at right sweeping up.* Wellcome Collection

65 STOCKTON & DARLINGTON RAILWAY

GEORGE STEPHENSON – COUNTY DURHAM, ENGLAND – 1825

Opening in 1825, the Stockton & Darlington Railway was the first in the world to use steam power to propel both passenger and freight trains.

As the Durham coalfield developed, so the existing means of transporting the mined coal – by packhorse – to the ports through which it was shipped southwards became increasingly inadequate. One of the key outlets for this traffic was Stockton-on-Tees in County Durham, where authorities had invested in improvements to ensure that the River Tees downstream was capable of handling increased traffic and were keen to promote the increased use of the port for the shipment of coal.

Proposals for the construction of a canal and later a railway linking Stockton with Darlington and the coalfield to the west first surfaced in the late 18th century, but it was not until 19 April 1821 that the Act permitting the construction of the Stockton & Darlington Railway received its Royal Assent. The key promoter of the line was the Quaker and Darlington-based woollen manufacturer Edward Pease (1767–1858) who, with the banker Jonathan Backhouse (1779–1842), were the major backers of the new railway.

George Stephenson (1781–1848), appointed engineer for the line, had previously been employed at Killingworth Colliery – where he had worked for Nicholas Wood (1795–1865), another of the key figures in the promotion of the new railway. It was while working at Killingworth that Stephenson designed his first steam locomotive – *Blücher* – in 1814. Although

◀ *George Stephenson was one of the most influential engineers of the Industrial Revolution. Involved in both civil and mechanical engineering projects, he is regarded as the 'Father of Railways' based upon his work on lines such as the Stockton & Darlington and the Liverpool & Manchester. It was his advocacy of the 4ft 8½in gauge that saw it ultimately adopted as standard gauge in Britain and throughout much of the world. Andrew Gray/ WikiCommons (CC BY-SA 3.0)*

◄ Locomotion No.1, *designed by George and Robert Stephenson and built by Robert Stephenson & Co., was the result of George Stephenson's earlier experience in constructing locomotives for the Killingworth Colliery.* WikiCommons

the mode of propulsion had not been specified in the original Act permitting the construction of the line, Stephenson was a staunch advocate of the use of steam when the line was completed and a new Act of 23 May 1823 permitted the use of 'loco-motives or moveable engines'. The same year saw Stephenson and Pease establish a locomotive building company – Robert Stephenson & Co. – in Newcastle upon Tyne and the first locomotive to be completed – for the Stockton & Darlington Railway – was *Locomotion No.1* in 1825. This locomotive was largely based upon knowledge that Stephenson had gained at Killingworth. With four driving wheels, it is believed that *Locomotion No.1* – known originally as *Active* – was the first locomotive constructed with coupling rods.

Delivered to the railway, *Locomotion No.1* hauled the first train over the Stockton & Darlington Railway on 27 September 1825. In so doing it became the first steam locomotive in the world to haul a train on a public railway. The first train, with George Stephenson himself as driver, comprised no fewer than 11 wagons of coal, a carriage called *Experiment* and 20 wagons full of passengers. *Locomotion No.1*

► *Replica of* Locomotion No.1 *on the Pockerley Waggonway, Beamish Museum.* Peter from Lincoln/WikiCommons (CC BY 2.0)

remained in use on the railway – despite being increasingly outdated – until 1841. It was then used briefly as a stationary pumping engine before being restored to railway use in 1846. A decade later, having at one stage survived the threat of scrapping, the locomotive was restored, being formally preserved the following year. Today, *Locomotion No.1* is part of the National Collection and is on display in Darlington; a working replica can be seen at Beamish.

66 ERIE CANAL

DEWITT CLINTON – ALBANY TO BUFFALO, USA – 1825

The largest and most ambitious civil-engineering project attempted in the USA at the time, the Erie Canal connected New York to the Great Lakes via the Hudson River and opened up the way to the West for settlement and commerce.

The Erie Canal was the vision of Governor DeWitt Clinton of New York to open up commerce and business into the interior of the USA and to secure New York's position as the most important port on the eastern seaboard. Turned down for funding by the Federal government, Clinton managed to secure $7 million from the New York State Legislature for the project and was named commissioner.

The canal started at Buffalo on the eastern shore of Lake Erie, ran through the Mohawk Valley gap in the Appalachian Mountains, then on to the Hudson River – in total 584km. It was 12m wide at the top, tapering to 8.5m

at the bottom, and 1.2m deep with a towpath alongside. It had to rise over 152m in elevation, requiring 83 stone locks and 18 aqueducts.

Construction began on 4 July 1816 on the middle flat stretch from Rome to Utica, New York State. It was so successful that income from the traffic covered this part of the cost. The next section was trickier, needing 50 locks. The builders and engineers had no experience of the technology and had to learn on the job.

The biggest challenge on the western side was the Niagara Escarpment, a 23m rock ridge. Engineer Nathan B. Roberts provided the answer in the form of a series of ten locks, five levels with two locks side by side. The next section required a 5km-long, 9m-deep cut blasted through the plateau. The Eastern Section from Brockport to Albany was opened on 10 September 1823. It ran for 50km and required 27 locks to overcome a series of natural rapids.

Small companies were contracted to dig short sections of the canal to spread the work

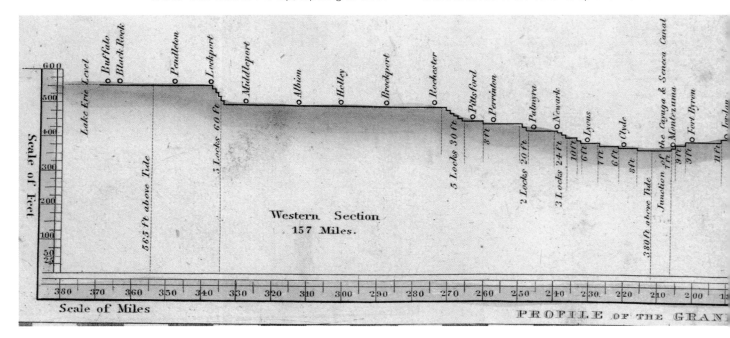

► *An aerial view of the Lockport section of the Erie Canal, looking southwest. Here, between 1817 and 1825, builders constructed a double set of five combined locks (centre of the photo) each with a lift of about 3.5m to take the canal over the 18m rise of the Niagara Escarpment.* Library of Congress

around. Each firm took responsibility for hiring people, horses and equipment, then supervising and paying their workers. Overall some 9,000 men were employed to build the canal, the majority of them Native Americans together with many Irish navvies. Labour was cheap and plentiful and sadly there were many fatalities during construction.

On 26 October 1825 the canal officially opened, two years ahead of schedule, with the Great Celebration when Governor Clinton journeyed the length of the canal on the *Seneca Chief*. He started in Buffalo to volleys of cannon shots all along the way.

Canal boats transported around 30 tons of produce, primarily grain and wool, but also whiskey and meat: the cost of transportation between Buffalo and New York fell from $100 per ton to around $25 per ton, and kept falling further. Within three years the tolls charged on the canal had completely repaid the state loan (plus interest) and financed several branch canals.

▼ *This contemporary drawing of the profile of the Erie Canal graphically describes the challenges of the project. In particular, it illustrates the dramatic Niagara Escarpment that required 27 locks to bring vessels up and down the 177km Eastern Section.* WikiCommons

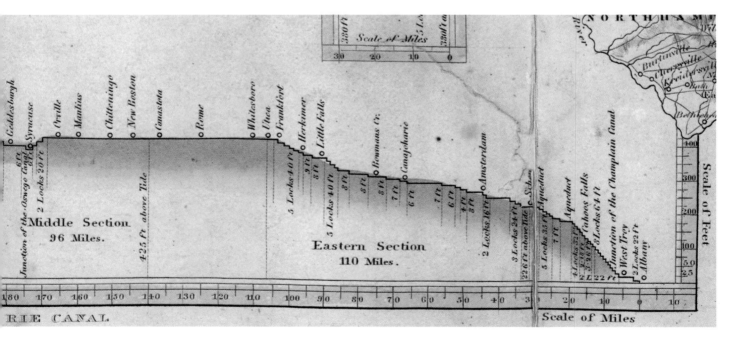

67 MENAI SUSPENSION BRIDGE
THOMAS TELFORD – ISLE OF ANGLESEY, WALES – 1826

The construction of the bridge across the fast-flowing Menai Strait improved transport links between Britain and Ireland following the Act of Union and demonstrated an alternative to traditional bridge designs.

▼ *The suspension bridge across the Menai Strait was designed by Thomas Telford as part of a major programme to link together mainland Britain and Ireland, following the Act of Union. This view of the bridge was taken in about 1860 from the Anglesey side.* Francis Bedford/The Marjorie and Leonard Vernon Collection, gift of The Annenberg Foundation, acquired from Carol Vernon and Robert Turbin/LACMA

In 1801 – as a result of insurrection and the threat of French involvement in Ireland – an Act of Union between Great Britain and Ireland was passed. This resulted in the end of the Irish parliament and instead representation for Irish MPs and peers in the House of Commons and House of Lords respectively at Westminster. One of the consequences of this legislation was an awareness that communication needed to be drastically improved between London and Holyhead. The noted engineer Thomas Telford (1757–1834) was employed to take up the challenge to find a solution. Much of the upgrade involved the improvement of what is now the A5 from

London through Lichfield and Shrewsbury and on towards Anglesey.

There was, however, a major challenge to reaching Holyhead. There needed to be a fixed crossing over the Menai Strait to Anglesey itself. Telford's design had to overcome a number of obstacles. The nature of the strait – with its fast-flowing water and high banks – presented a major problem as construction of a conventional bridge with piers would have been impractical. This was also ruled out as the Admiralty was concerned that the deck of the bridge had to be high enough above the water to permit Royal Navy warships to pass underneath. As a consequence, Telford proposed – and got approval for – the construction of a suspension bridge. Work started on construction in 1819 and was completed in time for the bridge to be officially opened on 30 January 1826.

The new bridge was built by William Hazledine (1763–1840) with the ironwork for construction being produced at his Shrewsbury ironworks. The main span of the bridge is approached from the Anglesey side by three stone arches; on the Bangor side, by four. These lead to the main towers that provide support to the suspension cables for the main span. The towers – like the rest of the stonework – are constructed from Penmon

limestone. They are hollow in construction with internal cross-walls. Early plans for these towers show that the upper levels were originally intended to be constructed in iron and that the bridge also included a central walkway between the two roadways. However, the towers were in the end constructed entirely in stone and the notion of a central walkway was abandoned.

The central span is 176m long and, at high water, 30.5m above the tide. The main span is supported by 16 iron chains, each of 935 links. Between 26 April and 9 June 1825, teams of labourers and a pulley system strung the chains across the strait. Each chain is 522.3m long and embedded into the bedrock on either side.

Almost from the moment of the bridge's opening, the wooden roadway was a problem and required modification. The resident engineer – William Alexander Provis (1792–1870) – undertook the replacement of the existing roadway in the late 1830s. His version was to survive until 1893, when it was replaced by steel. In the 20th century, the bridge, which is now Grade I listed, has undergone further modification, most notably between 1938 and 1941 when the existing iron chains were replaced by chains made of steel.

The bridge remains in use to this day, although heavier traffic now uses the Britannia Bridge (see page 154).

▲ *Almost two centuries after its completion, the Menai suspension bridge – despite replacement and strengthening work – is still recognisable in this view looking from the mainland towards Anglesey.* Nilfanion/WikiCommons (CC BY-SA 4.0)

68 OLDEST SURVIVING CAMERA PHOTOGRAPH

JOSEPH NIÉPCE – FRANCE – *c*1826

The first recorded image marks the starting point for a chain of developments that ended up with today's motion and still photographic technologies.

▶ *A portrait of Joseph Nicéphore Niépce after a painting by Leonard-François Berger, which is in the Musée Denon, Chalon-sur-Saône, France.* WikiCommons

▼ *This is an enhanced version of Niépce's 'View from the Window at Le Gras', taken with a camera obscura in either 1826 or 1827. It is the earliest surviving photograph of a real-world view.* WikiCommons

The oldest surviving photograph came about as the result of a device called the camera obscura. This was a popular drawing aid of the early 1800s, and was basically a box with a pinhole on one side. When suitably positioned, it would project an image of the scene before it onto a nearby screen. When the French inventor Joseph Nicéphore Niépce (1765–1833) realised that he lacked the necessary ability to record the image using conventional artistic techniques, he looked for another method of doing so.

What he came up with was labelled 'heliography'. This involved projecting images onto primitive light-sensitive silver chloride-coated paper. He faced two major problems: one was that the paper would darken as soon as it was brought into the light, losing the image entirely; the other was that he could record only a negative image – where there should be light the image showed dark, and vice versa.

The process that Niépce ended up with used a thin coating of liquid made from a solution of Bitumen of Judea dissolved in lavender oil. This was then smeared over a suitable background, such as a plate of glass, a flat stone or a sheet of metal. Once dry, the subject to be copied – such as an engraving – was laid over the top and the resulting assembly placed in direct sunlight. The dark areas of the engraving would mask the sunlight, whereas the light areas would not. As exposure to the sun hardened the bitumen – and so reduced its solubility – Niépce was able to use a solvent to dissolve the soft areas, leaving an image that could then be etched with acid or used for lithography.

The first images that Niépce produced in this way did not survive his attempts to make prints from them. In around 1827, though, he photographed the view from a window onto a sheet of bitumen-coated pewter. This is the oldest known photograph in the world, and from it all modern photography derives.

69 ROYAL WILLIAM VICTUALLING YARD

JOHN RENNIE – PLYMOUTH, ENGLAND – 1826–35

The Royal William was built to accommodate all Royal Navy supply needs at one central, purpose-built, secure location near the important Devonport Dockyard.

Supplying the vast Royal Navy with food, drinks and provisions was a massive job entailing commerce with numerous contractors and suppliers and all the accompanying possibilities for fraud and corruption. Logistically it was a nightmare.

The problem was addressed by the Victualling Board, the body charged with providing some 140,000 men (in 1810) with sufficient supplies to keep them fit and able to defend the Crown's interests, wherever in the world they might be. In 1823, the board commissioned a vast facility in Plymouth (and a later one in Portsmouth in 1827) to consolidate all the victualling requirements of the Royal Navy at one purpose-built location, where contracts for food and provisions for the Navy for ships and personnel overseas could be ordered, made and supplied. This cut civilian interference and profiteering to a minimum and greatly facilitated the quality and production of goods and services to the Royal Navy.

Designed by architect and civil engineer Sir John Rennie (1794–1874), the complex was built between 1826 and 1835 in the Plymouth suburb of Stonehouse. Located beside the deep waters of Plymouth Sound, the yard was named after King William IV (previously Duke of Clarence), who laid the foundation stone. It occupies around 6.5 hectares, 4 hectares of which were reclaimed from the sea. It comprises a collection of handsome, spacious buildings constructed from Plymouth limestone and Dartmoor granite and arranged on a symmetrical grid layout. This was designed to maximise efficient manufacture, storage and distribution of goods, which could be delivered straight onto vessels moored at the quays. The area includes a tidal basin, wharf walls, a brewery, a cooperage, a mill, a bakery, storehouses, a slaughterhouse, offices and residences for senior naval officers, and a guardhouse.

Each building is named: Clarence (1829–31) was the first section to be built and was the liquid store with a floor each for spirits, vinegar and beer. As a result much of the structure, roof doors, windows and so on were made of iron to be as inflammable as possible.

Up to 100 bullocks a day were slaughtered in the Slaughterhouse (1830–31), then the meat was salted into wooden barrels. The Mills Bakery (1830–34) was the bread and biscuit factory. Melville (c1828–32) was the administrative heart of the yard and the major storehouse for food, equipment and clothing.

The Brewhouse (1830–31) never actually worked as a brewery – it stood empty until 1885 when it became a repair workshop and rum store. Cooperage (c1826–32) was where 100 coopers made barrels and kegs to store liquid and produce. Guardhouse (1830–31) was for the naval police force.

In all, the yard cost about £2 million – much more than the 1825 Parliamentary estimate of £291,512 – but it served the Navy until 1992.

▼ *Contemporary view of the Royal William Victualling Yard at dusk. The magnificent complex of buildings has recently been sympathetically converted into a collection of private apartments, shops and restaurants.* Michael Chapman/ WikiCommons (CC BY-SA 4.0)

70 BIRMINGHAM CANAL NAVIGATION AQUEDUCT

THOMAS TELFORD – BIRMINGHAM, ENGLAND – *c*1828

Thomas Telford's technical innovation was his use of the iron-trough method to allow aqueducts to straddle much greater widths than those made using traditional masonry methods. He went on to refine this engineering breakthrough with his later, more celebrated work on suspension and road bridges.

During the Industrial Revolution, England's Midlands was laced together by an increasingly elaborate network of canals dug across the hilly landscape. Controlling the water within these canals required a complicated system of locks and pumps to ensure that the canals never ran dry.

The Engine Arm (or Birmingham Feeder) of the Old Main Line is a short canal designed and built by Thomas Telford to convey water from the Rotten Park Reservoir (now Edgbaston

► *The aqueduct was the first to use the iron-trough method for carrying water across another stretch of water – an idea much copied by subsequent canal engineers.* Oosoom/ WikiCommons (CC BY-SA 3.0)

Reservoir) over the BCN (Birmingham Canals Navigation) New Main Line to the adjacent and parallel Old Main Line. Part of the Engine Arm's purpose was to connect the reservoir with the Smethwick Engine that pumped water from the Birmingham level of the canal up to Smethwick Summit (Wolverhampton level) of the Old Main Line. The new cut allowed canal boats to carry coal directly to the pump engine.

Built and designed by Thomas Telford (1757–1834) around 1828, the aqueduct took the canal over the deep cutting of the new BCN Main Line at Smethwick. For this he devised a 15.9m-long, cast-iron trough-span aqueduct, supported by a single arch with five ribs, each consisting of four sections with bolted joints. Oriented northwest to southeast, it had an 2.4m-wide trough supported on three inner arches with some of the weight taken by the outer arches by way of vertical plates. Additionally, diagonal struts are cross-braced onto stone and brick abutments to counter the outer thrust caused by the weight of water in the trough.

▲ *Soberly painted in brown and cream with touches of red, the elegant, Gothic, cast-iron arches of the Birmingham Canal aqueduct make it one of the highlights of the Midlands' canal system. The ironwork was cast at the Horseley Ironworks, the foundry established by Aaron Manby, the man who became famous for constructing the first iron steamboat (see p.103).* Oosoom/WikiCommons (CC BY-SA 3.0)

The parallel towpaths on the east and west sides are each 1.3m wide, and supported by a row of elegant, fluted Gothic cast-iron columns and pointed arches. All the numerous castings required were manufactured at the nearby Horseley Ironworks in Tipton. The eastern side of the towpath has paved brickwork with raised strip footholds to give the towing horses grip.

On the northwest end of the aqueduct is an associated hump-backed roving bridge that allows the towpath to cross the entrance to the aqueduct. This is oriented northeast to southwest (90 degrees to the bridge) and built of blue engineering brick with stone copings and rusticated stone-arch details.

71 STOURBRIDGE LION
FOSTER, RASTRICK & CO. – STOURBRIDGE, ENGLAND – 1829

Two locomotives – *Stourbridge Lion* and *Rocket* (see opposite) – demonstrate the difference in locomotive design at the end of the 1820s. *Stourbridge Lion* – the USA's first steam loco – harks back to traditional designs.

▶ *A plaque at Honesdale – the 'birthplace of the American Railroad' – where the* Stourbridge Lion *first ran. It was named for the lion's face on the front and the place it was built – Stourbridge in the Midlands in Great Britain.* Ben Dalton/WikiCommons (CC BY 2.0)

By the third decade of the 19th century, steam locomotive development was moving on apace and two locomotives produced towards the end of the period demonstrated the advances – the *Stourbridge Lion* and the *Rocket*, both of which were completed in 1829.

The *Stourbridge Lion* – the first steam locomotive to operate in the USA – was the epitome of the traditional design. It possessed a single flue boiler with no separate smokebox nor firebox. The vertically acting piston rods were operated by two grasshopper beams – one for each cylinder placed above the boiler.

The locomotive was built at the workshops of Foster, Rastrick & Co., owned in Stourbridge by James Foster (1786–1853) and John Urpeth Rastrick (1780–1856). It was one of three locomotives supplied by the company to the Delaware & Hudson Canal Co., one of the earliest railroads in the USA, and was shipped across the Atlantic in parts prior to reassembly at the West Point Foundry in New York. It first ran officially on test on 8 August 1829. However, the weight of the locomotive – 7.2 tons (as opposed to the stipulated 4 tons) – and the poor track (iron strips on wood rather than solid iron) meant that it proved unsuitable for use on the line. An attempt to sell the Stourbridge-built locomotives failed and they were gradually robbed of their parts.

Eventually, only the boiler of *Stourbridge Lion* survived, in use at a foundry until its eventual preservation. It now forms part of the collection held by the Smithsonian Institution based in Washington, D.C. A number of replicas – including one by the Delaware & Hudson itself – have been constructed.

◀ *In 1916, artist Clyde O. DeLand (1872–1947) painted* The First locomotive *remembering the* Stourbridge Lion. *Library of Congress*

72 ***ROCKET***

ROBERT STEPHENSON – NEWCASTLE, ENGLAND – 1829

Stephenson's *Rocket* is significantly different to the *Stourbridge Lion* and heralds the way to the future with its 0-2-2 wheel arrangement, near-horizontal cylinders and improved multi-tube boiler.

▶ *George Stephenson was born in 1781 at Wylam, in Northumberland. The house where he was born is now a museum owned by the National Trust. Erected by the Institute of Mechanical Engineers and unveiled in 1929, a plaque on the exterior of the house pays tribute to him and to his success in the Rainhill Trials.* Tony Hisgett/WikiCommons (CC BY 2.0)

▼ *Although the Liverpool & Manchester Railway was engineered by George Stephenson, the* Rocket *was the work of both George and his son Robert (1803–59). With his father primarily involved in the actual construction of the Liverpool & Manchester Railway, it was left to Robert to undertake the detailed design work and oversee construction in Newcastle upon Tyne. Prior to the celebration to mark the 150th anniversary of the Liverpool & Manchester Railway in 1980, a working replica of the* Rocket *was built. This – and an earlier replica – now form part of the National Railway Museum Collection.* Tony Hisgett/WikiCommons (CC BY 2.0)

Completed for competition in the Rainhill Trials – the exercise by which the Liverpool & Manchester Railways would determine its locomotive needs – the *Rocket* was designed by Robert Stephenson (1803–59) and built at the Forth Street Works in Newcastle upon Tyne. The rules of the Rainhill Trials – which laid down maximum weights based upon the number of axles, for example – helped to determine the locomotive's design. Stephenson decided to construct a two-axle locomotive. This meant that the finished locomotive had to weigh less than 4.5 tons. Stephenson reasoned that the winning design must be both light and fast.

The *Rocket* was revolutionary for a number of reasons. Although several of its features had appeared earlier in other locomotives, this was the first time some of them had been used together, setting the pattern for future steam-locomotive design. In order to keep the weight down, Stephenson adopted the 0-2-2 wheel formation, saving the weight of the coupling rods. This arrangement allowed the single driving wheels to carry the bulk of the weight, which improved adhesion, while the cylinders, historically placed vertically, were placed closer to the horizontal. This minimised any swaying motion. The pistons were connected directly to the driving wheels and the locomotive used an innovative multi-tube boiler. This had no fewer than 25 copper fire-tubes, which increased significantly the locomotive's efficiency compared with earlier machines run on single-tube boilers. A separate firebox and a blastpipe also improved efficiency. The former ensured the optimum use of the coke to create heat; the latter saw the more efficient use of the exhaust steam to create a vacuum – it was only practical through the use of a multi-tube boiler.

Although modified during its life, *Rocket* was preserved on withdrawal in 1862 and is now part of the National Collection. A number of replicas have been built over the years.

73 LIVERPOOL & MANCHESTER RAILWAY

LANCASHIRE, ENGLAND – 1830

The Liverpool & Manchester was the first railway designed to link major centres of population and the first intended for operation by mechanical power alone.

As industrialisation and urbanisation developed, so the existing, water-based transport infrastructure became increasingly incapable of dealing with the resulting traffic. One consequence of this was that businessmen who were forced to rely upon the canals believed that the canal proprietors, lacking competition, were making excessive profits. Manchester was developing as a major centre of the cotton industry, but the canal companies were charging problematic rates for moving the vast quantities of raw material. Meanwhile, Liverpool – developing into the country's major port – was affected by the high cost of moving food to feed its growing population.

The solution – proposed by John Kennedy (1769–1855), a prominent Manchester cotton spinner, and Joseph Sanders (1785–1860), a Liverpool-based corn merchant – was a railway line between the two towns. In this they were influenced by William James (1771–1837), who conceived the idea that railways could form a transport network beyond the existing wagonways. When he died in 1837, James' obituary commented that he 'may with truth be considered as the father of the railway system, as he surveyed numerous lines at his own expense at a time when such an innovation was generally ridiculed'.

The Liverpool & Manchester Railway Co. was established on 20 May 1824. George Stephenson (1781–1848) was appointed engineer (see page 112) and he undertook additional survey work to supplement the work that James had completed earlier. However, the parliamentary procedure showed Stephenson's work to be inadequate and George Rennie (1791–1866) and his brother John (1794–1874; see p.119) replaced him. Charles Blacker Vignoles (1793–1875) was appointed surveyor. The railway was authorised by Act on 5 May 1826 but, unable to agree terms with the Rennie brothers for the line's construction, the railway reappointed Stephenson, with his assistant Joseph Locke (1805–60) as engineer.

▼ *Map of the Liverpool and Manchester Railway showing the line as surveyed by George Stephenson.* Science & Society Picture Library/SSPL/Getty Images

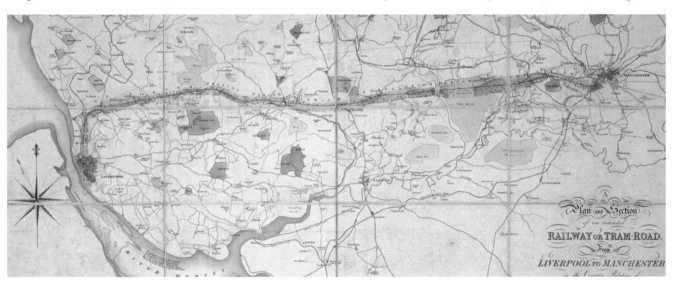

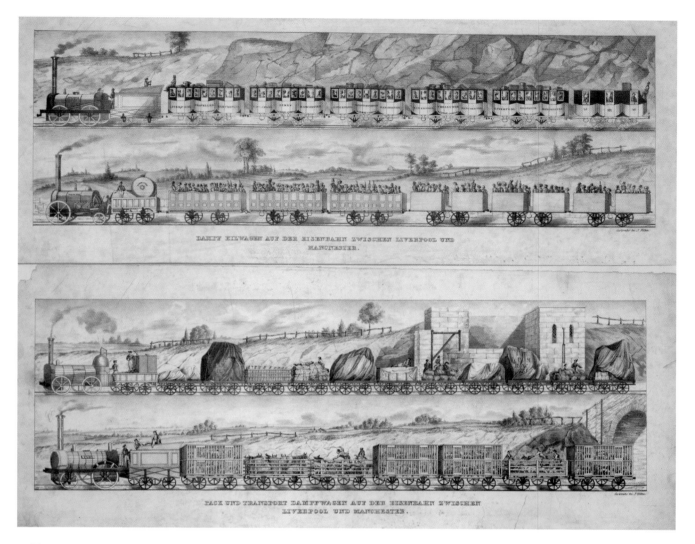

DAMPF EILWAGEN AUF DER EISENBAHN ZWISCHEN LIVERPOOL UND MANCHESTER.

PACK UND TRANSPORT DAMPFWAGEN AUF DER EISENBAHN ZWISCHEN LIVERPOOL UND MANCHESTER.

The construction of the line was a major enterprise. The 7.6km crossing of Chat Moss – an area of bog – proved challenging and required considerable ingenuity to complete. In all the route included 64 bridges and viaducts, including one near Manchester that was pioneering in its use of a cast-iron girder to support the railway line. The line opened ceremonially on 15 September 1830.

The Liverpool & Manchester Railway pioneered the notion of a mechanically powered passenger route. The line was also the first to be completed as double track throughout (most earlier lines were largely single track); and it was the first to use signalling. This was achieved by stationing policemen – hence the nickname for signalmen of 'Bobbies' – alongside the line, positioned about 1.6km apart. They used their arms to indicate whether it was safe for the

▲ *The Liverpool & Manchester Railway was perceived as the world's first railway connecting major urban centres. Although conceived originally as a means by which the businessmen of Manchester and Liverpool might provide competition for the existing canals (and so lower rates for freight traffic), passenger traffic was also developed from the line's opening. This contemporary engraving shows to good effect a number of the locomotives designed to compete with Stephenson's* Rocket *at the Rainhill Trials (see p.123) as well as the wide variety of rolling stock for which the line was designed to cater.* Library of Congress

locomotive to proceed. (The system of signalling was gradually modified by the introduction of coloured flags.)

The greatest significance of the line, however, was that it demonstrated that railways could be profitable and compete effectively against the canals. It was this line's viability that resulted in the railway age.

74 LAWNMOWER

EDWIN BUDDING – GLOUCESTERSHIRE, ENGLAND – 1830

The lawnmower revolutionised outdoor sporting events. Before the lawnmower's invention, cutting grass was a skilled, but time-consuming, expensive and laborious task with a hand scythe. Mown grass greatly improved the development of sports as well as the manageability of domestic gardens.

▼ *An early lawnmower. Before its invention only wealthy individuals and institutions could afford to have manicured lawns and fields. The arrival of the lawnmower allowed organised sports to expand and become popular pastimes open to the masses.* Clive Streeter/Getty images

Edwin Beard Budding (1796–1846) was a mechanic and freelance engineer and inventor, building and repairing machinery in the textile mills around the Stroud valleys in Gloucestershire. In the late 1820s, he developed a pistol that was allegedly better than Sam Colt's revolver of 1835, and he helped George Lister improve the carding machine in 1843. He also designed new versions of the lathe and spanner. However, watching the rotary cross-cutting machines at Brimscombe Mill that were used to cut the nap off woollen cloth and leave a smooth finish, he came up with a similar principle for cutting large areas of grass – such as at sports fields and in country gardens.

Budding went into partnership with businessman John Ferrabee, whose role was to pay the cost of development, obtain the letters of patent, sort out the manufacturing details, market the machine and sell licences to other manufacturers to build their own lawnmowers.

The lawnmower was patented in August 1830 when it was described as 'a new combination and application of machinery for the purpose of cropping or shearing the vegetable surface of lawns, grass-plats and pleasure grounds'. Manufactured at their facility at Thrupp near Stroud, the first lawnmower was a 48.25cm machine on a wrought-iron frame pushed using a pair of handles. The main, heavier, land roller at the back drove gears to transfer the drive to the knives on the cutting cylinder. The ratio was 16:1. The second, smaller forward roller was adjustable to alter the height of the cut. The action of the blades threw the grass cuttings into a large box on the front. An extra handle was soon added at the front to pull the machine along.

One of the very first machines was bought and used in the grounds of Regent's Park Zoological Gardens in central London. The other was purchased by the Oxford Colleges. The instructions were easy to follow, ' ... take hold of the handles, as in driving a barrow, ... push the machine steadily forward along the greensward, without lifting the handles, but rather exerting a moderate pressure downwards ...'

In 1832 Ransomes of Ipswich bought the first manufacturing licence from Budding and began selling the machines – ordinary people could at last cut the grass at their homes. According to Ransomes' advertising, 'The machine is so easy to manage that persons unpractised in the art of Mowing, may cut the Grass on Lawns, and Bowling Greens with ease.'

Amazingly, the principle of the original Budding lawnmower is largely unchanged, even in modern mowers.

75 JONES FALLS DAM
REDPATH AND MCKAY – ONTARIO, CANADA – 1832

When it was completed in 1832, Jones Falls Dam was the tallest dam in North America and across the British Empire. It was constructed to arch into the headwaters, which then 'locked' the entire structure in shape, holding back the enormous volume of water behind it.

Jones Falls Dam was built on the Rideau Canal in Ontario, Canada as part of the Jones Falls Lockstation complex. The dam was designed to hold back the enormous volume of water that ran into the 1.6km-long series of rapids flowing from Sand Lake. These waters

▼ *Jones Falls Dam as seen today – its novel design has clearly stood the test of time. As an unexpected byproduct, the fertile soil at the base of the southern face shelters and protects a unique plant community moistened by seepage from the dam.* Dennis Nazarenko/WikiCommons (CC BY 2.0)

feed into Jones Falls, then tumble around 11.9m down into the Pottawatomi River and ultimately flow into Owen Sound Bay.

The dam was designed and built by John Redpath and Thomas McKay under the general supervision of Lt-Col John By, the supervisor of the Rideau Canal. To control the flow from Sand Lake behind the dam during construction, two sluices were kept open to drain the waters: one on the east side near the base and the other on the west side about 6.1m above the base. When the dam was almost complete, Sand Lake was drained, the sluices were blocked and then the dam was rapidly finished to its full height as the lake refilled.

The dam is built of large sandstone blocks that were hauled 9.6km from the quarry at Elgin. About 110m long and built in a gentle curve, it is 18.25m high and 8.25m thick at the base. No mortar or cement holds it together: instead, the huge stone blocks were set in the form of a giant arch so that the pressure of the water held back by the dam pushes the precision-cut blocks together – using the same principle as that of a stone arch. The foundations were dug 2.5m into the lake bed and supplemented with an underwater rubble and mud slope that extended 38.7m upstream to reduce the pressure against the dam.

Some 260 men worked on the dam at its peak, 40 of them stonemasons, cutting and dressing the quarried sandstone. Disaster struck in summer 1828 when malaria killed dozens of men and incapacitated many others for weeks at a time. The malaria returned, though not so badly, each summer.

With the dam in place, the rapids were greatly reduced in volume. An adjustable weir was also constructed to control the level of the lake. Thanks to its shape, the dam is known as the Whispering Dam because someone standing at the top at one end can easily speak to someone standing in the same position at the other.

76 NEW YORK & HARLEM RAILROAD

JOHN STEPHENSON – NEW YORK, USA – 1832

The first ever street railway (or tramway), the New York & Harlem, was to influence the development of public transport throughout the world.

▼ *As horse tramways developed during the mid- to late 19th century, technology improved. Grooved track-and-flanged wheels were adopted and the track was incorporated into the regular road surface, thus making for a smoother ride for all road users. This 1880 image shows the original Grand Central Depot opened in 1871 (the existing structure was built 1903–13). In front are horse-drawn 'streetcars'.* New York Public Library Digital Collection

Alongside the great changes in technology that arose as a result of the inventions of the 18th century, there was an equal – and perhaps more profound – revolution in the way that society was organised: the trend towards urbanisation. Large cities had existed before the industrial age, but the vast majority of people lived in the countryside and their lives depended upon the annual cycle of agriculture. Factories required the concentration of the workforce in a single community and the development of transport was essential

▶ *The New York & Harlem Railroad is widely recognised as the world's first street railway and originally opened in 1832. The track was relatively crude with an 'L' section, while the wheels on the tram, unlike conventional railway vehicles, were unflanged.* Barry Cross Collection/ Online Transport Archive

both to move these people and supply them with food.

The world's first street railway was designed by the Irish-born John Stephenson (1809–93) and constructed on the eastern side of Manhattan Island, New York. The New York & Harlem Railroad, designed to provide a link between New York City and Harlem, was originally incorporated on 25 April 1831. The first section of the line – from Prince Street northwards to 14th Street along Bowery – was opened on 26 November 1832. Over the next 20 years, the line was extended both northwards and southwards so that it eventually stretched from Broadway to Chatham Four Corners. Initially the line was horse operated but, in 1837, the company introduced steam locomotives on the section north of 23rd Street. However, in 1854 restrictions on the use of steam resulted in the horse operation resuming as far north as 42nd Street.

It was another American – George Francis Train (1829–1904) – who brought the concept of the street tramway to the UK. His pioneering line was in Birkenhead, in Merseyside, where services commenced on 30 August 1860. He followed these up with three lines in London, the first of which – from Marble Arch to Portchester Terrace – opened on 23 March 1861. All were horse operated. Despite the popularity of the services, which were better than the existing horse buses, the method of construction – with rails standing proud of the existing road surface – caused problems for other road users. As a result these early pioneering services – along with a further one in Darlington that operated from 1862 to 1865 – were short-lived.

Although a handful of tramways were constructed in the UK during the late 1860s, when the promoters sought (and got) private Acts to permit construction, it was only after the passing of the Tramways Act of 1870

that a framework to permit widespread construction was established.

Although most horse tramways disappeared more than a century ago, it is still possible to experience one: Douglas, on the Isle of Man, possesses the world's last commercially operational horse tramway.

▼ *The horse-drawn railroad at 42nd Street between Fifth and Sixth Avenues* **c***1900.* New York Public Library Digital Collection

77 LONDON & GREENWICH RAILWAY
LONDON, ENGLAND – 1836

The London & Greenwich Railway can lay claim to a number of firsts: it was the first passenger railway to service London; it was the first elevated railway in the world; and it had the first signal-controlled railway junction.

▼ A fascinating panoramic view of the River Thames looking towards the east shows the first wet docks constructed to serve London. Of particular note are the views of the railway viaducts – among the earliest to serve London. On the north bank of the Thames is the London & Blackwall while on the south the London & Greenwich can be seen heading towards Greenwich, with Sir Christopher Wren's Royal Naval College prominent. The London & Croydon forms a junction with the London & Greenwich in the foreground. Wellcome Collection

The first railway to serve London, the London & Greenwich was a pioneering line in a number of ways. The railway was initially promoted by Colonel George Thomas Landmann (1779–1854), of the Royal Engineers, and George Walter (1790–1854), an entrepreneur and backer of early railways. A company for the line's construction was established on 25 November 1831, while an Act of Parliament dated 17 May 1833 gave powers for the route based on a survey undertaken in 1832 by Francis John William Thomas Giles (c1787–1847).

The London & Greenwich was the first railway to be promoted primarily for passenger traffic; its terminus at Tooley Street – later called London Bridge – was designed to provide a link

for passengers travelling from Greenwich into the City of London.

In order to avoid level crossings, the new railway was designed to be built as an elevated line. The principle of a railway crossing a single-span bridge had been established more than a century earlier at the Causey Arch on the Tanfield Railway, which was completed during 1725 and 1726 by Ralph Wood. The new railway was built on a viaduct of 878 brick arches running for a distance of some 6km. The line's main contractor was Hugh McIntosh (1768–1840), who later went on to be one of the contractors working on Brunel's Great Western Railway.

The first section of the line opened from Deptford (which today can claim to be the oldest passenger station in London) to Spa Road on 8 February 1836. It was extended to Bermondsey Street in October 1836 and to Tooley Street on 14 December 1836. From Deptford east, the line reached a temporary terminus at Church Row in Greenwich on 24 December 1838 and, following completion of the bridge at Deptford Creek, through to the station at Greenwich on 12 April 1840. Although the line carried a significant number of passengers, it was not the financial success that its original promoters anticipated. However, additional income was generated through the leasing out of the spaces under the arches.

The London & Greenwich was part of a more ambitious scheme – not completed until later – to provide a route through Gravesend and on to Dover. While the line from Greenwich eastwards through Woolwich opened in 1878, the opening of a second railway from Tooley Street – the London & Croydon (authorised by Act of 12 June 1835), which formed a junction with the existing London & Greenwich at Corbetts Lane – resulted in the existing line achieving another railway 'first': the erection of the first signal designed to control a junction. This was installed for the opening of the London & Croydon on 5 June 1839 and comprised a disc, operated by a pointsman, which indicated how the points were set: if the disc showed white (or a red light at night), the points were set for Croydon; if only the edge was visible (or a white light at night), the road was set for Greenwich.

▲ *The London & Greenwich was the first passenger railway to serve London. Built on a viaduct in order to avoid level crossings, among the buildings that the new railway passed was the church of St James's in Bermondsey. When the railway opened, the church itself was relatively new, having been consecrated only in 1829.* WikiCommons

▼ *The original terminus of the London & Greenwich was replaced in 1878 by a new through station constructed for the South Eastern Railway when the original line was extended through to the east. The new station was slightly to the east of the original terminus. The design of the new station owed much to the neo-Venetian design of the first station, by George Smith (1782–1869), which was completed in 1840. The newer station – illustrated here – is now Grade II listed.* Tom Betts/WikiCommons

78 TELEGRAPH

COOKE AND WHEATSTONE – LONDON, ENGLAND – 1838

The first commercial telegraph allowed simple messages to be sent quickly over distance and was initially adopted by railway operators sending instructions up and down the line.

In May 1837 inventor and entrepreneur William Fothergill Cooke (1806–1879) and another English scientist and academic, Charles Wheatstone (1802–1875), in a joint venture, patented a telegraph system that used electrical impulses to move needles on a board to point to letters of the alphabet that in turn indicated a code. More or fewer needles could be used

▶ *Back view of the five-needle telegraph patented by Charles Wheatstone and William Fothergill Cooke in 1837. The system used a diamond grid of 20 letters (the six missing letters had to be omitted from messages) with five needles arranged across the middle. The deflection of any two needles to the left or right pointed to specific letters.* Getty Images

◄ *William Fothergill Cooke (left) and Charles Wheatstone were inventors who built on the work of the European pioneers of telegraphy to patent a practical telegraph system.* Alamy; Wellcome Collection

depending on the number of characters and codes required.

On 25 July 1837 a four-needle system was demonstrated on a section of line between Euston and Camden Town to the directors of the newly opened rail line between London and Birmingham. The signal indicated that some carriages were ready for hauling up the incline between the two stations. However, the invention – the telegraph – was not taken up.

The first successful commercial use of the telegraph – a five-needle, six-wire system – was installed on the Great Western Railway over a 21km stretch between Paddington Station and West Drayton in 1838. Initially the cables were installed underground through a steel conduit, but they soon decayed in the unsuitable conditions and were replaced by uninsulated wires hung between poles. When in 1843 the line was extended to Slough, Berkshire, a one-needle, two-wire system was installed.

The telegraph sent electrical impulses to a diamond-shaped grid where pairs of needles would point to a letter. The number of wires corresponded to the number of needles, which in turn determined the number of characters that could be encoded. Early models used five magnetic needles. These were turned slightly left or right by electromagnetic induction generated by an energising winding. The sending operator toggled pairs of buttons that connected coils to the positive and negative ends of the battery. At the receiving end, the wires were all commoned together and the current energised the same pair of needles that then pointed to the relevant letter.

By 1867 a sixth wire was added that indicated the new addition of numerals. Soon, compound codes were included for railway operator controls (such as, 'Halt').

As the technology improved, all the new railways being built out from London installed an electric telegraph. The Blackwall Tunnel Railway had a Cooke and Wheatstone telegraph when it opened in 1840 and others were put in place right across the country.

To run an efficient railway system, trains had to run to a timetable, a near impossibility before the electric telegraph as different parts of the country ran on their own clocks based on sunrise and sunset. The electric telegraph established Greenwich Mean Time across the country and news could be transmitted widely and quickly. And not just for the good of the railways: in 1845 a murder suspect, John Tawell, was apprehended after a telegraph was sent from Slough to Paddington requesting his arrest when he got off the train. The incident became a sensation and alerted the general public to the wonders of the telegraph.

79

SS *GREAT WESTERN*
ISAMBARD KINGDOM BRUNEL – BRISTOL, ENGLAND – 1838

The SS *Great Western* was an oak-hulled paddle-wheel steamship designed to cross the Atlantic. She was the largest passenger ship in the world from 1837 to 1839, and her introduction stimulated a period of huge change in the maritime world.

In the mid-1830s there was a great desire for an efficient transatlantic shipping service between Bristol and New York. Although many people felt that large ships would not

▼ *This engraving shows the oak-hulled steamship SS* Great Western *'Taking her departure from Bristol for New York', captained by Lieutenant Commander James Hosken. The largest passenger ship in the world from 1837 to 1839, she was built to cross the Atlantic between Great Britain and the USA.* New York Public Library Digital Collection

work well in this role, Isambard Kingdom Brunel (1806–59) had worked out that this was not true. His calculations showed that the amount of cargo they could carry increased with the cube of their dimensions, while the drag they experienced increased only as the square of their dimensions. In other words, large ships were more fuel efficient, a matter that was of paramount importance on long voyages.

In 1836 Brunel got together with a group of friends and formed the Great Western Steamship Company to establish such a service. The first ship they built was the SS *Great Western*, a side-wheel paddle steamer with an oak hull that was reinforced with iron straps. It had four masts that could carry auxiliary sails – these were there not only to provide extra propulsion in a straight line, but to help stabilise the ship in rough seas. The problem being that if a paddle came out of the water, it not only impeded progress, but also imposed extra stresses on the engines.

Built in Bristol by Patterson & Mercer, the SS *Great Western* was launched on 19 July 1837. From Bristol she sailed to London for the two Maudslay, Sons & Field side-lever steam engines to be fitted – these were powerful for their day, together producing 750hp.

When she went into service, she was the biggest steamship in existence, but a year later the title was lost to a rival called the *British Queen*. In all she completed 45 crossings between Bristol and New York, and was still able to make record Blue Riband trips up until 1843. After eight years, however, her owners went out of business and she was sold to the Royal Mail Steam Packet Company in 1847, who used her for sailings to and from the West Indies. She then served as a troop ship during the Crimean War after which, in 1856, she was scrapped at Castles' Yard, Millbank, on the Thames in London.

80 VULCANISED RUBBER

CHARLES GOODYEAR – WOBURN, USA – 1839

In the mid-1800s, the rubber industry had been on the point of collapse owing to the product's inherent weaknesses where high or low temperatures were concerned. Charles Goodyear's discovery of how to vulcanise rubber revolutionised matters overnight.

▶ *Charles Goodyear's vulcanisation process for rubber made it possible to manufacture objects that had previously been either too expensive or impossible to make. This photograph taken inside one of his factories shows a few such items, including a boat as well as several pontoons.* Library of Congress

◀ *An engraving of Charles Goodyear by W.G. Jackman. Goodyear didn't fare well financially, with the British inventor Thomas Hancock patenting his idea in the UK and a technicality cancelling his French patent.* Library of Congress

Rubber, which is made from latex, the sap produced by the *Hevea* tree, was long known to the indigenous peoples of South America. While it had been used by the Olmec and Aztec cultures for centuries for a variety of purposes, it suffered from a major deficiency in its natural state. It was unstable when exposed to either heat (when it melted) or cold (when it cracked). A lot of investors had staked fortunes on the industry, but once the issues with stability became widely known, it teetered on the point of collapse.

After finding out about the shortcomings, Charles Goodyear (1800–60) became obsessed with finding a solution. He began a long series of experiments that put his family into serious debt. He moved several times to wherever there appeared to be a chance of funding for his research, and spent countless hours both at home and in debtors' prison mixing up various concoctions in his seemingly endless quest to solve the problems of rubber. He came up with what appeared to be various answers, but all of them failed for one reason or another.

It was only when he was working at the Eagle India Rubber Company in Woburn, MA, that he made his breakthrough. He accidentally put some sulphur into some rubber that was on a hot stove – and vulcanisation was born. It took him several years to get the process right, but eventually it was good enough and he was granted a patent in 1844. This saved the rubber industry from its imminent demise, and all manner of new products were manufactured as a result, including such diverse items as life jackets, vehicle tyres, pencil erasers, gloves, and rubber balls. The method was then licensed to other manufacturers, which brought in a lot of money. Unfortunately, despite all this success, he lost all of his fortune – and more – through fighting numerous patent infringement lawsuits. When he died at the age of 59 in 1860, he was seriously in debt.

81 PEDAL BICYCLE

KIRKPATRICK MACMILLAN – DUMFRIES, SCOTLAND – 1839

The origins of the bicycle date back many centuries, but the mechanically propelled version was not invented until 1839. Since then, cycling has gone through many phases, and today it remains as popular as ever.

▶ *The plaque to Kirkpatrick Macmillan at the Courthill Smithy, Keir Mill, Penpont, Dumfries & Galloway.* Rosser1954/ WikiCommons (CC BY-SA 4.0)

▼ *The first true bicycle – powered by moving pedals – was probably built by a Scottish blacksmith called Kirkpatrick Macmillan, in 1839. It is thought that the example shown here was a copy of his creation made about 20 years later. It had a large rear wheel in an attempt to make up for the lack of gearing.* Getty Images

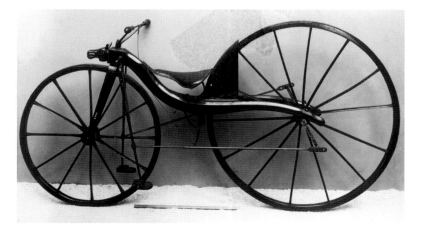

Exactly who invented the bicycle will forever be a matter of contention. Literature of the time suggests that Robert Hooke (1635–1703) may have had the first idea for a bicycle in the late 1600s, as he was said to have careered around the streets of London on some kind of wheeled conveyance. The first fully documented ancestor of the modern bicycle was the German *Draisine*, which dates back to 1817. However, it was a primitive affair that was powered by moving the feet along the ground. It sparked a number of copies, though, known as 'hobby horses'. The first true bicycle – powered by turning pedals – was probably built by a Scottish blacksmith called Kirkpatrick Macmillan (1812–78), in 1839. Again, the specific details are somewhat murky, but he is generally credited with the invention.

The bicycle that he constructed is said to have been made of wood with iron-rimmed wooden wheels. The front one was steerable, while the larger, rear one, was driven from the pedals by connecting rods. The bicycle was very heavy indeed, and the pedals, sited near the front, were operated back and forth in a reciprocating manner.

There was a report in a Glasgow newspaper in 1842 of a 'velocipede of ingenious design' knocking over a pedestrian, with the pilot of the machine being fined five shillings for causing the accident. A descendant of Macmillan's, called Johnston, claimed that Macmillan was the rider mentioned in the paper – although again it's not clear whether or not this was true.

Macmillan never patented his idea and it is claimed that another inventor called Gavin Dalzell, of Lesmahagow in Lanarkshire, Scotland, beginning in 1846, copied Macmillan's machine so extensively that for a long time he was considered to be the actual creator of the bicycle. These days, though, Macmillan's name has been resurrected, and he currently enjoys the claim for himself.

82 STEAM HAMMER

FRANÇOIS BOURDON/JAMES NASMYTH – FRANCE/SCOTLAND – 1840

The steam hammer was invented in response to a serious engineering dilemma: how to manufacture the ever-bigger machinery that was being created on a daily basis. It solved the problem efficiently and cheaply.

James Hall Nasmyth (1808–90) was a Scottish engineer and inventor who was told by one of the engineers in charge of the construction of Brunel's SS *Great Britain* (see page 144) that there wasn't a hammer in existence big enough to forge the paddle shaft. At that stage the ship was intended to be paddle-driven, and so needed appropriately sized driveshafts. Intrigued by the issue, Nasmyth drew up a steam version in his sketch book, dated 24 November 1839. The immediate need went away, however, when shortly afterwards, Brunel saw the improvements provided by screw propellers, and did away with the paddle concept altogether.

Around the same time a French engineer called François Bourdon was also working on the same subject. He built the first working steam hammer in the world at the Schneider & Cie works at Le Creusot in 1840. Weighing in at 2,500kg, it was a substantial machine that lifted the hammer to 2m. It was only when Nasmyth saw it in action in 1842 that he went home, patented his design, and built his own version. Unsurprisingly, a dispute later broke out between the two men as to who had priority over the invention.

Despite the ill feeling that it generated, the hammer itself was a commercial success. It not only made it possible to manufacture bigger components, but it improved quality and reduced costs by over half. One of the unique features for the time was that the force of the hammer blow could be finely controlled by the operator – this allowed for a lot more precision throughout the operation. In particular, large forgings like ships' anchors could be made far more efficiently as they no longer needed to be fabricated from lots of small pieces that were then welded together. Within a very short time, Nasmyth's hammers could be found in workshops all over the country, and by 1856 an incredible 490 hammers had been manufactured and delivered to clients right across Europe to Russia, and as far beyond Europe as India and Australia.

▶ *The steam hammer – seen here in action in a painting by its inventor, James Nasmyth – not only permitted the manufacture of far larger items than had previously been possible, but also significantly reduced production costs. As a result, it was a great commercial success and was sold right across the world.* Getty Images 463983697

83 SCREW-PILE LIGHTHOUSES
ALEXANDER MITCHELL – LANCASHIRE, ENGLAND – 1840–41

Screw-pile lighthouses were designed to stand on piles over soft, shifting bases such as sand or mud. They are literally screwed into the bed of the river or sea.

▼ Diagram showing the Maplin Sands Lighthouse. WikiCommons

Screw-pile lighthouses were the inspiration of Alexander Mitchell (1780–1868), a blind Irish engineer. They were locked deep into the sea or river bed using piles secured with broad-bladed screws. The lighthouses did not tower above tall waves, and were intended for use in areas such as river deltas, where the warning light does not need to be visible from a great distance. In 1833 Alexander Mitchell and his son patented the design for their wrought-iron screw-pile lighthouse.

In 1838 construction of the first screw-pile lighthouse began at the mouth of the River Thames on the dangerous mudflats of the northern bank off Foulness Island. It was designed by James Walker, consultant lighthouse engineer for Trinity House, using Mitchell's screw-pile concept.

Known as Maplin Sands Lighthouse, it had nine screw piles, arranged as a central pile and eight surrounding piles, supported by a lattice of cast-iron struts. On this was built an octagonal wooden platform that had living accommodation for a principal keeper and two assistants. They shared a single, three-bunk bedroom, a living room, a kitchen-washroom and a storeroom. Above this was a bright red light tower, standing 21m tall. Its fixed lamp was placed 11m high and was visible for 16km. The lighthouse also had a flagpole on one side and a 136kg fog-warning bell on the seaward side that gave one stroke every ten seconds in fog. The lighthouse was first lit in 1841, but eventually the unforgiving tides and flow of the

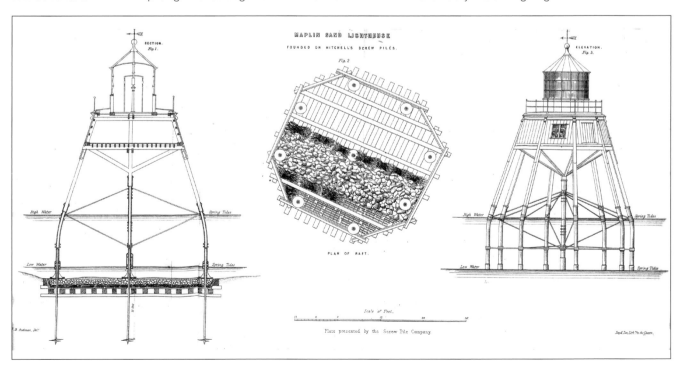

Thames undermined it and it was swept away in 1932.

Although construction of the Maplin Sands Lighthouse started before any other, it was not the first screw-pile lighthouse to operate. That accolade goes to the Wyre Light in Fleetwood, Lancashire, at the edge of Morecambe Bay, in 1840. Standing 3.7km off shore, the Wyre Light sits on the North Wharf Bank, a line of sandbanks that marks the Fleetwood Channel into the Wyre estuary. Started after the Maplin Sands Lighthouse, it was much quicker to build – and so to light. It was also designed by Mitchell and built by his firm.

This light had seven wrought-iron piles arranged as a central pillar and six surrounding pillars to provide a hexagonal platform. Each pile was 4.8m long with a cast-iron screw base 1m across embedded into the sands. The platform held a two-storey building to house the keeper. The main hexagonal room was 6.7m in diameter and 2.7m high and divided into two – the living quarter with fireplace and a dormitory.

Construction started in 1839 and the lantern lit on 6 June 1840. It shone 9.4m above high tide level and could be seen for 12.8km. The fog bell could be heard for 3.2km. Unfortunately, the lighthouse caught fire in 1948 and it was not replaced.

▲▼ *The Screw-pile lighthouses have proved popular all over the world, particularly in locations where erosion or shifting sands create problems for traditional lighthouse construction.*
The Sand Key Lighthouse (Above), built on a reef that is intermittently covered by sand, was completed in 1853. It was in use, surviving hurricanes, until 2015. United States Navy, James Brooks
The Thomas Point Shoal Lighthouse (Below) on the Chesapeake Bay near Annapolis, Maryland was originally built from stone but erosion led to the erection of this screw-pile lighthouse. It was activated in 1877 and is still there today, unaltered save for automation in 1986. Shutterstock

84 CROTON DISTRIBUTING RESERVOIR

JOHN JERVIS AND JAMES RENWICK – NEW YORK, USA – 1842

The reservoir provided clean, safe drinking water on a city-wide scale for the first time.

Also known as the Murray Hill Reservoir, the Croton Distributing Reservoir was located from 40th to 42nd Street between Fifth and Sixth Avenues in Manhattan. It was built as the main source of clean and reliable drinking

▼ *The Croton Distributing Reservoir received, held and then distributed fresh water across New York. It was designed in imposing Egyptian Revival style and provided panoramic views across the city from its upper walkways.* New York Public Library Digital Collection/WikiCommons

water for 19th-century New Yorkers. Previously citizens relied on cisterns, natural springs, wells and collected rainwater. But in the 19th century, New York's population rapidly expanded and the existing polluted and unsanitary water supply caused outbreaks of diseases such as yellow fever and cholera. Clean, drinkable water became a priority. Additionally, the growing industries located in the city required large volumes of water in order to function.

The Croton Reservoir drew water from the Croton River Lake in northern Westchester County. Work on the project started in 1837 and took five years to complete. The aqueduct started with a 65km-long series of complicated

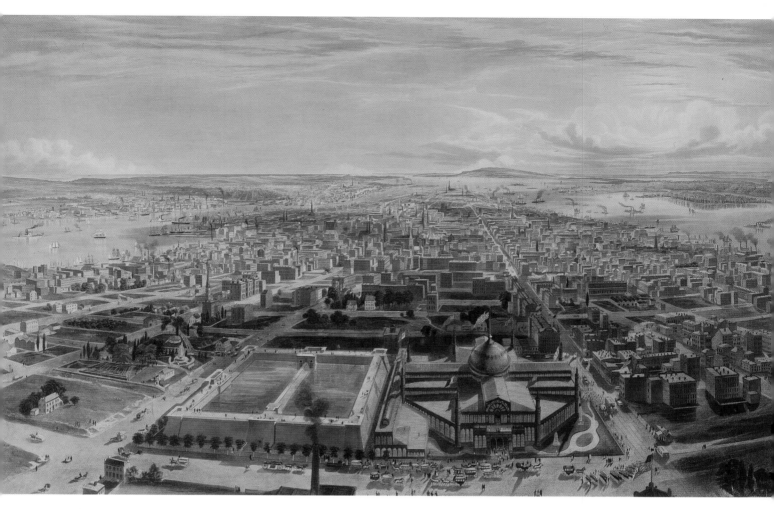

underground iron-pipe conduits encased in brick masonry, which flowed all the way to New York. The aqueduct spanned the Hudson River at 173rd Street using the High Bridge, then moved down the west side of Manhattan where it poured into the Croton Reservoir in the area then known as Yorkville.

This magnificent structure was designed by engineer John Jervis (1795–1885) and architect James Renwick (1815–95), in imposing Egyptian Revival style and was recorded as costing almost $500,000. The reservoir occupied a site some 1.62 hectares in area, and had 15.25m-high granite walls that were 7.5m thick. It was divided in the centre by a wall of granite 5.8m thick at the bottom and 1.2m thick at the top, splitting the water into two collection pools. The reservoir was 550m long and 255m wide and capable of holding a capacity of 75 million litres of water. It received millions of gallons of water every day, and this

▲ *This 1855 etching shows the twin holding tanks and the broad walkways around the perimeter and across the centre of the reservoir. It was used for only 35 years before it became obsolete.* New York Public Library Digital Collection/WikiCommons

was distributed across New York via a network of pipes some 275km in length.

Around the perimeter, heavy iron railings enclosed promenade walls some 6.1m wide. These provided magnificent views out over the city and panoramic views over Long Island and New Jersey. They were a popular spot to take the air.

Twenty thousand people witnessed the official opening on 4 July 1842. However, by 1877 the reservoir had become obsolete and there were calls for it to be demolished. *The New York Times* called it 'useless, a hideous object to the sight, and a blight upon the neighbourhood'. It was decommissioned in 1897 and soon demolished.

85 DALKEY ATMOSPHERIC RAILWAY
SAMUDA BROTHERS AND SAMUEL CLEGG – DUBLIN, IRELAND – 1843

In an era before the dominance of steam traction on the railways, the Dalkey Atmospheric Railway proved that there were alternative means of propulsion – even if the most extensive use of the invention proved one of Brunel's failures.

▼ *Following research and seeing the Dalkey line in action, Isambard Kingdom Brunel decided to adopt the principle for the South Devon Railway from Exeter to Plymouth. This illustration – reproduced in MacDermot's classic history of the Great Western Railway – shows the theory behind the operation of the atmospheric railway.* Author's Collection

While today history records that it was the steam engine that came to dominate the development of the railway age, to those contemporaries involved in the early railway schemes the dominance of steam was by no means a given. There were pioneers who investigated other means of propulsion and one of the most promising was the atmospheric railway. There were concerns that steam locomotives would be incapable of hauling trains up significant gradients and the atmospheric railway seemed to offer a viable alternative.

Opened in 1834, the Dublin & Kingstown was the first railway in Ireland. However, in order to extend the line to Dalkey – a distance

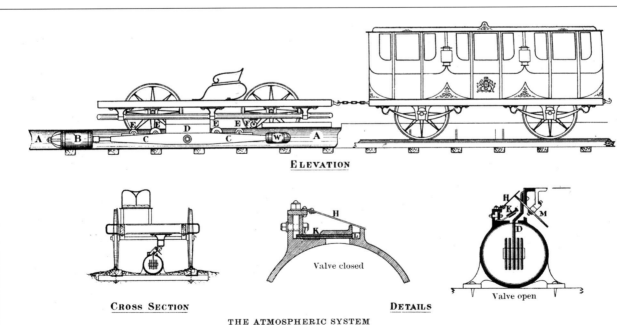

ELEVATION

CROSS SECTION

THE ATMOSPHERIC SYSTEM

DETAILS

Valve closed

Valve open

A.A. Continuous Pipe fixed between the rails.
B. Piston.
C.C. Iron Plates connected to the piston.
D. Plate connecting Apparatus to Carriage.

E. Metal Rollers to open the Continuous Valve.
F. Roller attached to Carriage for closing the Valve.
H. Weather Valve.[1]

K. Continuous Airtight Valve hinged at l.
L. Composition for sealing Valve.
M. Roller attached to Carriage for opening Weather Valve.[1]
W. Counterweight to Piston.

[1] These complications do not appear to have been in use on the South Devon Railway.

of just under 3km – the line needed to ascend at an average gradient of 1 in 110 (1 in 57 at its most severe) and it was decided to adopt – for the first time in the world – the atmospheric principle of operation espoused by Jacob Samuda (1811–44), his brother Joseph d'Aguilar Samuda (1813–85) and their colleague Samuel Clegg (1781–1861). They developed the concept – patented in 1838 – of a continuous cast-iron tube allied to a slot sealed by a leather flap. This was demonstrated over 800m of the West London Railway in 1840 and operated for about two years.

The Dalkey Atmospheric Railway was engineered by Charles Blacker Vignoles (1793–1875) and the contractor was William Dargan (1799–1867). The atmospheric equipment was supplied by the Samuda brothers and by Clegg. The cast-iron pipe was 380mm in diameter and was fitted only to the uphill line – the downhill trains relied on gravity. The vacuum to the pipe was supplied by a single 100hp steam engine located at Dalkey and achieved speeds of up to 74km/h. The tube terminated about 512m short of the Dalkey terminus and trains relied on their momentum to reach their destination.

The line opened officially on 29 March 1844, although services had operated since 19 August 1843. The new railway was visited by a number of contemporary railway engineers, most notably Isambard Kingdom Brunel (1806–59), who was seeking a solution for the construction of the South Devon Railway, in particular the heavily graded section between Newton Abbot and Plymouth.

Impressed by his visit, Brunel adopted the atmospheric principle for the South Devon, although it was not fully installed by the time the first section of the railway opened – from Exeter to Teignmouth – on 30 May 1846. It was not until 13 September 1847 that the first services using the new equipment operated. However, along the sea wall at Dawlish, the system proved problematic: a combination of sand and the corrosive effect of the sea water on the leather flap made the vacuum almost impossible to maintain – resulting in the decision to abandon its use. It last operated on 10 September 1848. Although the section west

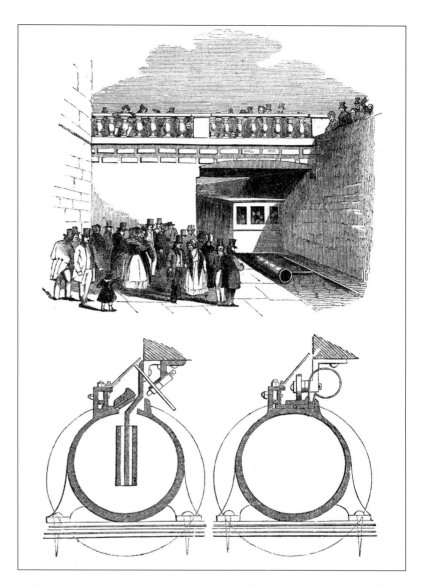

▲ *Early railway engineers were not convinced that steam would be sufficient to provide power to operate trains up gradients. One alternative was the atmospheric railway. The first practical use in the world of this short-lived technology was the Dalkey Atmospheric Railway in Dublin, which operated for a decade.* Getty Images

of Newton Abbot towards Totnes was partially fitted with the equipment, this was never actually used as the newest steam locomotives that ran there proved wholly capable of dealing with the ruling gradients without it.

Although the atmospheric principle was not to last long in England, in Ireland the Dalkey line proved a success. It was finally abandoned on 12 April 1854 when the line closed for reconstruction from 4ft 8½in to the Irish standard gauge of 5ft 3in.

86 SS *GREAT BRITAIN*
ISAMBARD KINGDOM BRUNEL – BRISTOL, ENGLAND – 1843

The SS *Great Britain* was another vessel designed for transatlantic service – she was not only the longest passenger ship in the world from 1843 to 1854, but also the first screw-driven iron steamer to sail across the Atlantic from Great Britain to the USA.

Isambard Kingdom Brunel's (1806–59) SS *Great Britain* was, when she was launched in 1843, by far the largest ship in existence. She remained the longest passenger vessel in the world right up until 1854. Built for the Great Western Steamship Company's transatlantic service by William Patterson (1795–1869), she plied between Bristol and New York. She was the first screw-driven iron steamer to cross the Atlantic, a feat that she achieved in 1845 in 14 days. Equipped with both a screw propulsion system and an iron hull – again, a first in a large ocean-going ship – she was 98m long and had a displacement of 3,674 tons. Motive power came from two twin-cylinder direct-acting engines – these had a bore of 220cm and a 1.83m stroke, developing 370kW. She also had five schooner-rigged masts and one that

▼ *An oil painting by Richard Ball Spencer of the* SS Great Britain *at Brunswick Wharf, the London and Blackwall Railway Company's terminus at Blackwall. Owned by the National Maritime Museum, Greenwich, London, it shows her flying the Royal Standard as well as the French and US flags. It is therefore likely that there was a royal visit to the ship at the time she was painted.* National Maritime Museum, Greenwich

was square-rigged to supply a secondary source of motion. There were four decks, which provided accommodation for the crew

of 120, as well as for 360 passengers who were given cabins, in addition to dining and promenade saloons. Later in her life, her capacity increased to 730 passengers and she could carry 1,200 tons of cargo.

Unfortunately for the Great Western Steamship Company, the ship building costs were astronomical: rather than the anticipated £70,000, she ended up costing £117,000. This put the company on a financial knife edge, and when a navigational error saw her run aground in Dundrum Bay, the costs of refloating her in 1846 forced the company to close. The SS *Great Britain* was sold for salvage in 1852, but was then repaired and went on to spend many years carrying thousands of passengers to Australia. In 1881, rather perversely, she was converted to sail, and then three years later she was sent to the Falkland Islands were she acted variously as a warehouse, a quarantine ship and a coal hulk, before being scuttled in 1937. There it seemed she was destined to remain. However, English businessman Sir Jack Hayward made a donation that paid for her to be towed back to the UK where she was restored – she is now on permanent display in a specially built outdoor museum at Bristol City Docks.

▲ *The SS* Great Britain *as she is today – moored at Bristol harbour where she is on display to the public after undergoing a full restoration.* Mattbuck/WikiCommons (CC BY-SA 4.0)

87 MORSE CODE

SAMUEL MORSE – MASSACHUSETTS, USA – 1844

The simple electronic system of dots and dashes devised by Samuel Morse (and the generally uncredited Alfred Vail) enabled near-instantaneous long-distance communication for the first time in history.

Early methods of communication over distance involved horses and messengers, or direct line-of-sight flag signals (such as semaphore) or fire beacons. With the Industrial Revolution and the discovery and harnessing of electricity, an entirely new form of communication became possible and many would-be inventors tried to work out a simple and effective method of harnessing the new technology to rapid communication.

The first step was made by Alessandro Volta (1745–1827) when he invented the battery that allowed the storage and control of electronic current. Then Hans Christian Oersted (1777–1851) showed how magnetism was allied to electricity. In the early 1830s Samuel Morse (1791–1872), a talented and successful portrait painter, became one of many inventors on both sides of the Atlantic working hard to discover the answer. And it was he who made the breakthrough via his experiments with electromagnetism. Working in collaboration with Professor Leonard Gale (1800–83) and Alfred Vail (1807–59), he eventually devised a single-circuit telegraph that sent an electronic signal along a wire to a receiver at the other end. The message was sent along the wire in a series of electronic pulses – long and short, known as dashes and dots. The specific bursts of dashes and dots each represented a letter and collectively built up into words.

◀ *A photograph of Samuel F.B. Morse. He made his name first as an artist before his involvement with telegraphy, brought on by the untimely demise of his wife. She fell ill, died and was buried before he could return to his home in New Haven from Washington where he was painting a portrait of Civil War hero Lafayette, who was touring the then 24 states of the Union. His wife's swift death led him to examine long-distance communication.* Library of Congress

Morse's telegraph required a length of wire, a line of poles to hold up the wire, a key, and a battery, receiver and operators at either end. After demonstrating their device to Congress in 1843, Morse and Vail received funding of $30,000 to set up a 60km-long experimental telegraph line between Washington, D.C., and Baltimore, Maryland. Their first message 'What hath God wrought!' was transmitted on 24 May 1844. Soon telegraphs were set up all over the USA and Europe.

Early Morse Code messages were written down on paper as marks by the receiving operator, who then translated them into letters and words. Soon, however, operators became so skilled that they could easily distinguish the letters themselves without having to

▲ An early Morse Code machine. Operators quickly became very adept at sending and reading the signals and the machines were provided with a louder click to make them clearer. Getty Images

decode them, so the clicking was made more pronounced for ease of understanding.

Morse received his patent for the telegraph in 1847 and immediately became embroiled in numerous legal claims from rivals and investors. In 1854 the US Supreme Court ruled in O'Reilly v. Morse that the latter had been the first to develop a workable telegraph.

By 1866 the transatlantic cable (see page 166) enabled rapid communications between the USA and Europe – using Morse Code.

88 HOWE SEWING MACHINE
ELIAS HOWE – CAMBRIDGE, USA – 1844

The Howe Sewing Machine was the first sewing machine to use a lockstitch design. From then onwards clothing, shoes and other textiles could be quickly and economically mass produced. The innovation created millions of new jobs worldwide.

Elias Howe (1819–67) started his professional life as an apprentice in a Massachusetts textile factory. Other people had already designed and even patented sewing machines, but they all used a chainstitch mechanism that looped into itself under the fabric. Sewn from a single thread, the seam held until the thread broke, when the entire line of stitches could unravel. No one had really solved the intrinsic problems of permanently fastening textiles together satisfactorily.

Howe's breakthrough was to work out how to lock the thread so that the stitches wouldn't undo. Called a lockstitch, his method used two threads, one above the fabric and the other worked from underneath and a needle with the eye at the point instead of at the base. When the needle plunged through the fabric, a shuttle carrying the second thread passed through the loop, catching

▶ *Many sewing machines preceded Howe's version, but his was the first to use the lockstitch – the use of two threads, one from above and the other from below.* New York Public Library Digital Collection/ WikiCommons

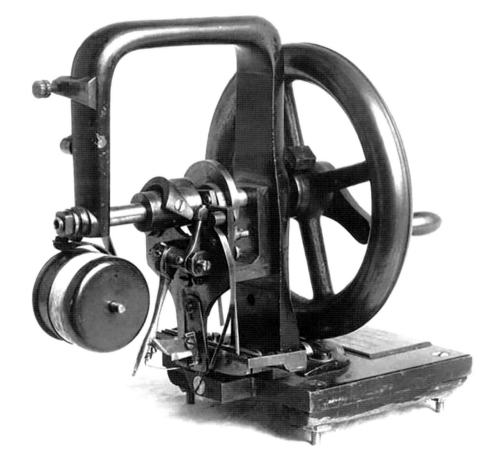

the upper thread. When the needle retreated upwards, the thread tightened and left a neat line of stitches that would not unravel even if the thread broke. This simple but efficient mechanism was thanks to three of Howe's innovations – automatic feed, a needle with an eye at the point (previous machines kept the eye at the top of the needle like a hand-stitching needle), and a shuttle that moved underneath the cloth that caught the loop of the thread above to form the lockstitch.

During tests it was found that hand-sewers averaged 23 stitches a minute, whereas the Howe Sewing Machine could make 640. A calico dress took around six and a half hours to make by hand but just under an hour by machine. The clothing industry was completely revolutionised.

In September 1846 Elias Howe patented his sewing machine in the USA as a 'process that uses thread from two different sources'. But he was almost immediately plunged into protracted litigation from 1849 to 1854 against Isaac Merritt Singer and fellow inventor Walter Hunt to establish his patent as they were blatantly producing a very similar lockstitch machine. Howe eventually won his court case and Singer had to give him a share in I.M. Singer & Co. profits and pay him considerable royalties, as did other copycat manufacturers. From this time Howe issued licences to other manufacturers permitting

them to use his patents and from 1856 he collected a royalty of $5 per machine sold in the USA through the patent combination. This made him a very rich man.

In 1863 Howe improved his hand-cranked sewing machine by designing a foot-press mechanism. His sewing machine became accessible financially for light manufacturing and home use.

▲ *The Howe Sewing Machine factory in Bridgeport, Connecticut, a stereo photograph taken in the 1870s.* New York Public Library Digital Collection/ WikiCommons

◀ *Elias Howe also patented a form of zipper in 1851, but didn't develop it and it was left to Gideon Sundback in the early years of the 20th century to perfect the design.* Library of Congress

89 ROYAL ALBERT DOCK
JESSE HARTLEY AND PHILIP HARDWICK – LIVERPOOL, ENGLAND – 1845

The completion of the Albert Dock in Liverpool marked the acme of Victorian dock development with the world's first non-combustible warehouses and the first use of hydraulically powered cranes.

▼ *Authorised by Act of Parliament in 1841, the Albert Dock was developed by Jesse Hartley in conjunction with the architect Philip Hardwick. The warehouses were conceived, designed and built to be as fire-resistant as possible and to that end were constructed in brick, stone and cast iron. A prominent feature of the structures are the massive cast-iron columns. Peter Waller*

During the 19th century the wet docks that served the port of Liverpool were significantly extended as trade – both the importation of raw materials and the export of finished goods – grew exponentially. Among the docks that were constructed during this period was the Albert Dock.

The key figure in the development of Liverpool's docks between 1824 and 1860, when he acted as Superintendent of the Concerns of the Dock Estate, was Jesse Hartley (1784–1860), who drew up the initial plans for a combined dock and warehouse system. The first dock to employ this concept, whereby ships were loaded and unloaded

▶ *One factor in the design of the dock was to make it as secure as possible in order to reduce pilferage.* Peter Waller

directly from a warehouse (which could reduce pilferage – a serious problem in docks and harbours) was St Katharine's Dock in London, which was opened in 1828.

However, Hartley's plans for Liverpool were more ambitious. Working with Philip Hardwick (1792–1870), who is perhaps now better known as the architect behind the much-mourned Euston Arch in London, Hartley was keen to create warehouses that were as fire-resistant as possible. After testing various structures, he decided to build the new warehouses using brick, cast iron, granite and sandstone. The results were to become the first buildings in Britain to be completed without any structural woodwork and the first non-combustible warehouses in the world.

Following the Act permitting construction of the Albert Dock, which received Royal Assent in 1841, building work commenced. The dock – albeit not wholly complete – was officially opened by Prince Albert in 1846. Integral to the design of the warehouses are the massive cast-iron columns. These are 4.5m high and 4m in circumference. One of the features of the new warehouses was the first use in the world of hydraulically powered cranes (see page 152).

The enclosed nature of the dock allied to its secure loading and unloading facilities meant that the Albert Dock was used for high-value goods – such as tobacco, silk and brandy – but its heyday was to be relatively short-lived.

The dock was designed to cater for sailing ships of up to 1,000 tons and larger vessels found it impossible to access the dock through its narrow entrance. By the late 19th century, sail was being replaced by steam and the size of vessels was growing significantly. By the 1920s commercial use of the dock itself had almost ceased, although the warehouses continued to be busy storing goods for onward movement.

Damage during World War II along with the financial problems of the dock's owner – the Mersey Docks and Harbour Board – seemed to presage that there was no future for the dock

and its warehouses (despite the complex having been Grade I listed in 1952) following their final closure in 1972. Plans for redevelopment came to nothing, but then the creation of the Merseyside Development Corporation in 1981 saw new life eventually brought to the complex. Today, the Albert Dock is home to attractions such as Tate Liverpool and the Merseyside Maritime Museum, making it one of the key visitor destinations on Liverpool's river frontage.

90 HYDRAULIC CRANE

WILLIAM ARMSTRONG – NEWCASTLE, ENGLAND – 1846

The hydraulic crane used water power to provide faster, greater lifting power at reduced cost. After its introduction it proved so popular – especially with dockyards and railways – that Armstrong's company manufactured an average of 100 cranes a year until 1900.

▼ *A schematic diagram showing the main components of Sir William Armstrong's hydraulic crane in both plan and elevation.*

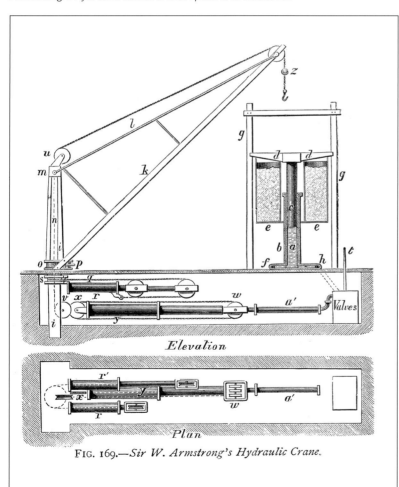

FIG. 169.—*Sir W. Armstrong's Hydraulic Crane.*

The hydraulic crane came about after English industrialist William George Armstrong (1810–1900) watched a waterwheel at work at a marble quarry while out fishing. Struck by its apparent inefficiency, he developed a rotary engine that was driven by water. This attracted little interest, so he went on to produce a piston-engined version. Buoyed by its performance, he looked around for something that it could power, and he settled on a hydraulic crane.

When it was decided to pipe water to the houses in the city of Newcastle from distant reservoirs, Armstrong was involved in the scheme. Realising that there would be excess water pressure, he proposed to the local corporation that he build a hydraulic crane on the quayside. He claimed that it would unload ships faster and more cheaply than conventional cranes, and after some discussion, the corporation agreed. The crane proved to be so good that the corporation added three more. Armstrong realised that he had discovered significant business potential, so he resigned his position as a lawyer and founded W.G. Armstrong & Company, located on 22,000m² of land that he bought next to the river at Elswick, near Newcastle, where he built a factory.

Once the factory was up and running, it wasn't long before he had manufactured and delivered hydraulic cranes to the Edinburgh and Northern Railways, as well as to Liverpool Docks (see page 150). He also got orders for hydraulic machinery to operate the dock gates in Grimsby. Once word of the efficacy of his equipment got out, the company was inundated with enquiries, and in 1850 it supplied 45 cranes. By 1852, this figure had risen to 75, and the company averaged 100 a year until 1900. It also became one of the major local employers, with 300 men on site in 1850. However, this was only the beginning – by 1863, there were over 3,800 employees.

Although cranes were the impetus for the company, it also took orders for bridges. In addition, Armstrong found ways to provide hydraulic power when there wasn't enough water pressure. Essentially, these were water towers and, later, the hydraulic accumulator.

▼ *The only Armstrong Mitchell hydraulic crane left in existence was installed at the Arsenale, Venice, Italy, between 1883 and 1885. Sadly, it is badly in need of restoration, without which there is serious risk of it collapsing.* Jakub Hałun/ WikiCommons (CC BY-SA 4.0)

91 | BRITANNIA BRIDGE
ROBERT STEPHENSON – MENAI STRAIT, WALES – 1846–50

Stephenson's great bridge across the Menai Strait provided, as well as an essential link in the improved communications between London and Dublin, proof that wrought iron, properly used, was suitable for use in major construction projects.

▼ *Designed by Robert Stephenson and built between 1846 and 1850 for the Chester & Holyhead Railway, the Britannia Bridge crosses the Menai Strait. This view, taken from the Anglesey side looking across the strait in the 1860s, shows the bridge as originally completed with the carved lions guarding each of the portals.* Francis Bedford/The Marjorie and Leonard Vernon Collection, gift of The Annenberg Foundation, acquired from Carol Vernon and Robert Turbin/LACMA

A generation after the construction of Telford's suspension bridge over the Menai Strait (see page 116), a second engineer, Robert Stephenson (1803–59), faced the same challenge as he sought to construct the Chester & Holyhead Railway's line across from the mainland to Anglesey.

As with the road bridge, Stephenson's bridge had to be constructed in such a way that fully rigged naval vessels could pass through the strait unimpeded. The Britannia Bridge – so named because the central stone-built tower stands on Britannia Rock in the middle of the channel – was to provide a double-track connection.

The bridge was formed of three masonry piers allied to two stone-built abutments, one on each side of the strait. The tracks were to

be carried by two parallel iron-made tubes constructed from wrought iron and riveted. The two central spans were to be 140m long with two outer spans being 70m long each. As a result the bridge's ironwork extended over some 461m with each tube weighing about 1,500 tons. At the time of the bridge's planning, the longest wrought-iron span ever constructed stretched for only 9.6m.

Recognising that he was seeking to stretch the material further than ever before, Stephenson sought advice from William Fairbairn (1789–1874), a distinguished engineer who had worked with Stephenson's father. Fairbairn enlisted the assistance of Eaton Hodgkinson (1789–1861) and, with the two of them, Stephenson was able to provide scientific support for his proposed design. Fairbairn had shown conclusively that through careful design it was possible both for the girders to support themselves as well as – more importantly – being able to carry the weight of a train as well.

Work commenced on the construction of

▲ Stephenson's bridge was to survive for more than a century until a serious fire in May 1970 compromised the original structure. This photograph shows the replacement bridge with its second level above the railway track for road traffic. WikiCommons

the bridge with the laying of the foundation stone on 10 April 1846. The final rivet was inserted – appropriately by Stephenson himself – on 5 March 1850 and railway services over the bridge commenced on 21 October 1850. The thoroughness of Fairbairn's research was demonstrated by the fact that the structure survived intact until 23 May 1970 – despite the increase in weight of locomotives and trains – when a fire fatally weakened the wrought-iron tubes. While the stone-built piers were reusable, the original tubes were not and so the spans were replaced by a new structure. The new bridge had two decks – a supported steel level for use by the railway and a higher (concrete) level for use by road traffic to relieve the traffic load on the older road bridge. This work was finally completed in 1980.

92

STEAM ENGINE
GEORGE H. CORLISS – PROVIDENCE, USA – 1849

Rotary valves much improved the thermal efficiency of stationary steam engines, which became considerably more economical than water power. Manufacturing industries using steam engines were no longer restricted by access to flowing water and millponds.

▼ *George H. Corliss – another person who received the soubriquet the 'American James Watt' – enormously improved the efficiency and mechanical detail of steam technology. Here a diagram of a Corliss-type valve gear shows the movement of high-pressure steam (in red) and low-pressure steam (in blue) moving through the cylinder. With each stroke the four valves alternate opening and closing, driving the piston back and forth.* Marcbela/WikiCommons (CC0)

George Henry Corliss (1817–88) improved the workings of steam engines with the introduction of independent rotary valves with variable valve timings. He invented a valve that allowed steam rapidly to pressurise each side of the piston, moving it back and forth before the steam had time to condense, which otherwise would draw heat from the engine, slow it down and cause it to lose power.

Corliss was granted a patent for his valve gear in March 1849. It describes a vertical cylinder beam engine with individual slide valves for admission and exhaust at each end of the valve. It also covered a wrist plate to move the valve motion from a single eccentric to the four engine valves and the use of trip valves with variable cutoffs. His engines proved on average to be around 30% more fuel efficient

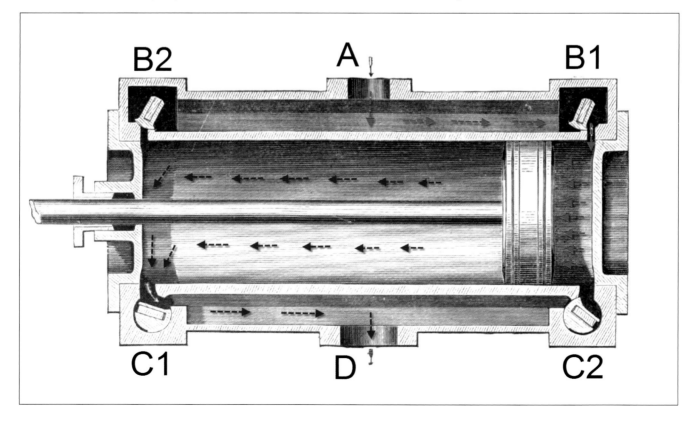

than earlier steam engines using a steam cutoff. The first Corliss engines had four separate inlet and exhaust valves for each cylinder. He also introduced springs to speed the opening and closing of the valves, meaning they could be independently controlled. This saved steam, and meant that for the first time the cylinders and valves were not subjected to continual drastic temperature changes and therefore loss of efficiency.

The Corliss valve gearing provided a more uniform speed and more reliable response to load changes, all of which made it ideal for lighter industrial use. The system was mostly used in stationary engines in factories and mills to provide mechanical power for line shafting via a series of pulleys, belts and gears.

Because of the delicacy of handling threads, the textile industry particularly welcomed the

benefits of its smooth running speed and swift response. The patented valve gear allowed machines to be precisely set and also its great power output could drive many machines at different ratios. These could be taken on and off line as required.

In 1876 the Corliss Centennial Engine powered almost all of the exhibits at the Centennial Exposition in Philadelphia. On show to the public in the Machinery Hall, it was one of the wonders of the show. It stood on a platform 17m across and was the largest steam engine in the world producing 1,400hp. Made of iron and steel, it towered 14m high with a flywheel 9m in diameter. It drove over 8km of belts, shafts and pulleys to other machines across the Hall. Eventually sold to work at the Pullman factory in Chicago, in 1910 it was sold for scrap at $8 per ton.

▲ *Corliss steam engine at the New England Wireless and Steam Museum, East Greenwich, Rhode Island. This is one of the very few remaining original engines built in the 1870s at Corliss' Providence Engine Works.* The-Daffodil/WikiCommons (CC BY-SA 4.0)

93 SALT'S MILL
TITUS SALT – SALTAIRE, ENGLAND – 1850

The entrepreneur Titus Salt created one of the largest textile mills of the age, but also oversaw the construction of one of the key 'model' communities of the industrial age – Saltaire – now a World Heritage Site.

▼ *The Grade II* listed Salt's Mill at Saltaire was designed by the Bradford-based architectural practice of Lockwood & Mawson and built between 1851 and 1853. The building's engineer was Sir William Fairbairn (1789–1874). The mill's lengthy main façade faces south onto the railway line. The architects' original design was rejected by Titus Salt as being 'not half large enough'. The building's exterior was in stone but the framework was constructed in brick and cast iron in order to minimise the risk of fire.* Peter Waller

Life for the mill workers in the late 18th and early 19th century was harsh – as recorded in Elizabeth Gaskell's first novel *Mary Barton* (1848), which portrayed the life of the workers of Manchester. However, there were a number of pioneering entrepreneurs that sought to ameliorate the mill workers' generally poor living and working conditions. One of these early pioneers was Robert Owen (1771–1858) at New Lanark, but another was the Yorkshire textile manufacturer Titus Salt (1803–76) – later to become Sir Titus when he was created a baronet in 1869. In 1833 Salt took over his father's business in Bradford and, following his decision to weave cloths from alpaca wool (which he'd discovered in Liverpool three years earlier), the company grew rapidly.

By the late 1840s Bradford was suffering pollution and, determined to build a new factory and consolidate all his businesses onto a single site as he 'did not like to be a party to increasing that already over-crowded borough', Salt purchased land to the north of Bradford, situated alongside the Leeds–Liverpool Canal and the Midland Railway.

The massive mills – the largest industrial complex in the world by floor area – when completed were the same length as St Paul's Cathedral in London. The architects for the complex were the Bradford-based Henry Francis Lockwood (1811–78) and William Mawson (1826–89) with Sir William Fairbairn (1789–1874) advising on the installation of the machinery. Formally opened on 20 September 1853 – to coincide with Salt's 50th birthday – the main spinning block was 165m long and comprised five storeys plus basement.

With the development of the mill came the construction of the model industrial village. Lockwood and Mawson were again employed as the architects, and they provided a grid of terraced housing, offering a quality of dwelling far superior to anything then available to the workers of the other textile mills in Bradford. Alongside the housing came a Congregational chapel, a school, a hospital, an institute and almshouses (but no pub), which looked after the residents from the cradle to the grave.

The motive for entrepreneurs like Salt is uncertain. Religion undoubtedly played a part in the desire to provide decent accommodation for their workers, but there was also a pragmatic view that a healthy workforce was likely to be more productive and more content.

Salt's Mill, which is now listed Grade II*, continued in operation until 1986 and, since then, has been restored for a number of other uses, including a gallery for paintings by the locally born artist David Hockney. The whole of the Saltaire complex is now recognised by UNICEF as a World Heritage Site.

► *Opposite Salt's Mill, and linked by a walkway, stands New Mill. Built in 1868, its chimney is modelled on the bell tower at Basilica di Santa Maria Gloriosa dei Frari in Venice, Italy.*
Tim Green

◄ *The statue of Sir Titus Salt (1803–76) in Roberts Park, Saltaire. The side panel illustrates the source of Salt's wealth – the alpaca. He first came across the wool of this South American mammal in 1836. Although not the first to try and use the wool, his experimentation with the fibre led to the development of a lustrous and fashionable cloth.*
Tim Green

94 THE CRYSTAL PALACE
JOSEPH PAXTON – LONDON, ENGLAND – 1850–51

The Crystal Palace – the showcase for the Great Exhibition of 1851 – was a masterpiece of architectural design using glass and steelwork in a manner never seen before on such a scale. Designed as a temporary structure to house exhibitors in a 92,000m² space, it was 564m long and 39m high.

The Great Exhibition of 1851 was intended to display the latest products from all over the world, but the problem was that no suitable building existed in which to house it. After a long and acrimonious process, out of more than 245 entries, Joseph Paxton (1803–65), a pre-eminent figure in British horticulture, was eventually chosen to design and build the solution. The commission in charge of mounting the Great Exhibition had been set up in January 1850 and included such famous names as Isambard Kingdom Brunel, Robert Stephenson, Charles Barry, Thomas Leverton Donaldson, the Duke of Buccleuch, the Earl of Ellesmere, and William Cubitt. As the costs of running the event were to be met by public subscription, the budget was limited, and Paxton's design was not only appropriate for the job, but was the cheapest on offer.

By the time the decision was made, Paxton

▼ *George Cruikshank's contemporary view of the opening ceremony on 1 May 1951, showing Archbishop of Canterbury John Bird Sumner offering up a prayer. Her Most Gracious Majesty Queen Victoria and His Royal Highness Prince Albert look on. Victoria dubbed it 'one of the greatest and most glorious days of our lives.'* Library of Congress

had only eight months in which to finish his designs, get all the components made and erect the building. That he managed to do so on time and within budget was a remarkable achievement in itself. Further to that, he had to make some last-minute changes, such as the addition of a high transept across the middle of the building in order to avoid having to fell some large elm trees that had become the bone of much public contention.

The building itself was constructed in Hyde Park, London, in a modular fashion from plate glass, wood and cast iron. While Paxton's methods were ground-breaking, it was the scale of the structure that was truly awe-inspiring – three times bigger than St Paul's Cathedral, the building covered 7 hectares. The design was based on the maximum size at which the plate glass could be made – each pane, manufactured by Chance Brothers of Smethwick, was 25.4cm wide and 124.5cm long. Making the entire structure from pieces of the same size saved huge amounts of money and drastically reduced the construction time.

The numbers involved are staggering. Engineering firm Fox, Henderson and Co. used more than 1,000 iron columns to support 2,224 trellis girders. To duct rainwater away there was 48km of guttering. At peak times there were more than 2,000 people working on the site.

The exhibition was a huge success, with more than 14,000 exhibitors from across the globe. Visitors were astonished that the building needed no lighting owing to the vast area of glass – the most ever seen in a building. As one of the primary stipulations was that the building had to be temporary, after the exhibition closed it was dismantled and re-erected in Sydenham, southeast London – it remained there until its destruction by fire in 1936.

▼ *An illustration from Dickinson's* Comprehensive Pictures of the Great Exhibition of 1851–1854 *of the magnificent Crystal Palace, as seen from the northeast.* WikiCommons

95 BESSEMER CONVERTER
HENRY BESSEMER – SHEFFIELD, ENGLAND – 1850

This converter enabled manufacturers to mass produce steel from molten pig iron.

Before Henry Bessemer (1813–98), the process to turn around 5 tons of iron into steel took a full day of stirring, heating and reheating. With the use of a Bessemer Converter, it took no more than 20 minutes.

Bessemer's ambition was to improve steel so that it could be used in quantity to make quality weapons, at that time for use in the Crimean War. The only contemporary steel was made in small quantities for cutlery and tools and was much too costly for larger objects like cannon.

At his bronze powder works in January 1855, Bessemer started investigating a way to produce large quantities of steel. He blew air through molten iron and after exposing it to extreme heat, produced a metal he called mild steel. He came up with a manufacturing process and sold the licence to four ironmasters, but their output proved faulty and he bought back the licences and returned to investigating the problem, spending tens of thousands of pounds in the process. He knew the iron needed a carefully controlled flow of air to burn off impurities but leave enough carbon in the metal. None of his licensees could make the process work, so Bessemer set up his own steel company.

In 1856 Bessemer took out his patent for the convertor and also read a paper to the British Association called 'On the manufacture of malleable iron and steel without fuel'.

Elsewhere in England metallurgist Robert Mushet (1811–91) discovered the solution was to burn off all the impurities and carbon and then reintroduce carbon and manganese in the form of specular pig iron, or *Spiegeleisen*. Imported from Germany, this was a compound of iron, manganese and carbon that absorbed oxygen. This greatly improved the quality of steel, especially its malleability. Unfortunately, Mushet was unable to keep up payment of the patent fee and it was bought by Bessemer.

The first commercial manufacture of steel started in Sheffield in 1858 when Bessemer opened his steel works using imported charcoal pig iron from Sweden and his large egg-shaped containers in which the iron was heated and melted.

In essence, iron is inserted into the convertor through a hole at the top. Heat is applied at the bottom. When the pig iron is molten, pressurised air is blown through to oxidise it and remove impurities. These escape as gas or turn to slag. Then, additives such as *Spiegeleisen* are added to the molten steel before it is ladled into moulds.

Bessemer Converters were usually operated in pairs: while one was being blown, the other was filled. They treated up to 30 tons of pig iron at a time. Additionally, the process used the impurities it removed to produce the heat needed for the process, meaning far less coal was used, greatly reducing the manufacturing costs of steel.

▼ *A Bessemer Converter dating from 1850 outside the former ironworks of Högbo, Sandviken Municipality, Sweden.* Jan Ainali/ WikiCommons (CC BY-SA 4.0)

96

HYPODERMIC SYRINGE

ALEXANDER WOOD – EDINBURGH, SCOTLAND – 1853

Alexander Wood pioneered the use of the hypodermic needle and syringe for the measured administration of intravenous drugs.

Even before the ancient Romans, hypodermic needles (of sorts) have been used to administer medicines such as enemas into the body. But it was not until the 19th century that the most significant breakthrough came. Scottish academic and Edinburgh physician Alexander Wood (1817–84) invented the first hypodermic (meaning 'beneath the skin') needle with a hollow tube and true syringe, anecdotally using the sting of the bee for inspiration. He initially developed it to administer morphia and opium. Wood's first patient was injected with liquefied morphine in 1853 to relieve pain, a procedure he described in *The Edinburgh Medical and Surgical Journal* (1855) in a paper entitled 'A New Method for Treating Neuralgia by the Direct Application of Opiates to Painful Points'.

At the same time in Lyon, France, veterinary surgeon Charles Pravaz (1791–1853) devised a similar pistol-driven syringe that became widely used in the medical profession under the name of the Pravaz Syringe. Both inventors used a metal barrel for the syringe and a fine, hollow metal needle. By 1866 the syringe barrel was made from glass so the dosage could be clearly seen as it was injected. Unfortunately, neither sterilisation nor appreciation of the likelihood of transferring disease from one patient to another was even suspected for many years. Unsterile needles caused severe skin abscesses in many users.

Wood used a syringe made by London instrument maker Daniel Ferguson who made a piston-equipped syringe that had a narrow, hollow-pointed needle; when the outer tube was rotated it aligned to an aperture in the needle. The plunger then pumped the medicine into the needle and on into the patient.

In its earliest years the hypodermic syringe was primarily used to administer morphine (an isolated variant of opium, synthesised in Germany in 1803), especially for the treatment of wounded soldiers in the American Civil War. It was popular among the combatants because it gave almost instantaneous pain relief. As a result some became addicted. A number of the soldiers' loved ones also took morphine to deal with their loss – again often resulting in addiction. In the second half of the 19th century, though, doctors believed that injected morphine did not pass through the stomach and digestive system, believing (wrongly) that, therefore, addiction wasn't possible. Many doctors were sceptical about the efficacy of the administration method, and it was only by the very late 1800s that syringes were widely used, although at that time there were still few drugs that could be injected.

In the USA it was possible to purchase syringes by mail order and it was even fashionable for some ladies to carry syringes around with them.

▼ *An English hypodermic syringe dating from around 1860. This one was designed by Coxeter & Son, London, specialist medical instrument makers.* Science Museum, London via Wellcome Collection

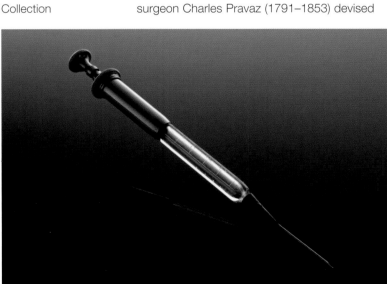

97 LADY ISABELLA WATERWHEEL
ROBERT CASEMENT – ISLE OF MAN, UK – 1854

This marrying of ancient and modern technology became an important step towards the modern water turbine.

In the 19th century the Isle of Man possessed an extensive mine system that contained rich mineral resources – silver, zinc, lead and copper – but no coal. The deeper the mines ran into the ground, the more

▶ This photograph shows the sheer size of the Lady Isabella Wheel. It took four years to construct and almost immediately became a symbol of the Isle of Man and an important tourist attraction in its own right. Lobster1/ WikiCommons (CC BY-SA 3.0)

◄ *An early photograph of the Lady Isabella waterwheel dating from between 1890 and 1910. At this date the wheel was continually working to pump water out of the extensive local mineshafts.* National Library of Ireland on The Commons/ WikiCommons

water filled the lower levels of the shafts. A modern steam engine was needed to pump out the water, but without coal to drive it (and importing being too expensive) mine-owners needed another solution. Self-taught engineer Robert Casement (1815–91), when presented with the problem, adapted ancient technology and applied it to the latest technology. He devised a system of channels to collect the plentiful water from the surrounding hillside streams of Glen Mooar and diverted it into a large cistern located above the wheel. From there, the water was piped across a bridge and into a tower above the great wheel where it fell into a series of 192 slatted wooden buckets, each capable of holding 11 litres of water, lining the rim of the wheel. The weight of the water turned the wheel.

After four years of construction, the 70-ton wheel started working in 1854. It was named after Lady Isabella Hope, the wife of the lieutenant governor of the Isle of Man at the time. It is also known as the Laxey Wheel, after the village it serves.

Contained within a large stone building, the waterwheel is 22.1m in diameter and has a circumference of 70m, with a width of 1.83m. It is traditionally painted bright red. It averages three revolutions a minute and is made from a combination of wooden wheel and rod, and cast-iron mechanical parts.

The crankshaft has a throw of 1.22m and connects to a counterweight and a very long rod that runs on small wheels to minimise friction. An enclosed pipe takes water to the top of the wheel, so the water flows up the tower as an inverted syphon. The water falls into the buckets fitted into the wheel to make the wheel rotate in a pitchback or reverse direction. Water runs to the pumping shaft where the 2.44m stroke is converted by a T-rocker into a pumping action to drive pumps 183m away at the mineshaft. The pumps could lift 142 litres of water a minute from almost 457m below ground. The extracted water was drained into the Laxey River.

At its peak the mine employed more than 600 miners, but it closed in 1929.

98 TRANSATLANTIC CABLE
CYRUS WEST FIELD – IRELAND TO CANADA – 1858

The world became a smaller place in 1858 when the first transatlantic cable started working. Previously messages could reach the distant shore only by ship – a passage that took a minimum of ten days.

In 1850 a working telegraph cable was laid between England and France and soon a project to link the Old and New Worlds became a reality. One of the leading lights of the project was Cyrus West Field (1819–92), an American businessman and financier. A proposed route was decided on – western Ireland to eastern Newfoundland – and a survey of the sea floor undertaken. To fund the project, shares were sold in the Atlantic Telegraph Company. Field made four attempts to lay a successful transatlantic cable – the first in 1857, two attempts in 1858 and another in 1865 – before the technological problems were solved and he achieved success in 1866.

For the attempt in June 1858, the cable was laid by the converted warships HMS

▶ *Hand-coloured lithograph dated 1861 showing the USS* Niagara *and HMS* Agamemnon *starting to lay the cable. The paying-out machinery used to drop the cable into the ocean is clearly visible at the stern of the* Niagara. Library of Congress

TO CYRUS W. FIELD, ESQ.

ATLANTIC TELE GRAPH POLKA.

"THE NIAGARA & AGAMEMNON COMMENCING TO LAY THE CABLE."

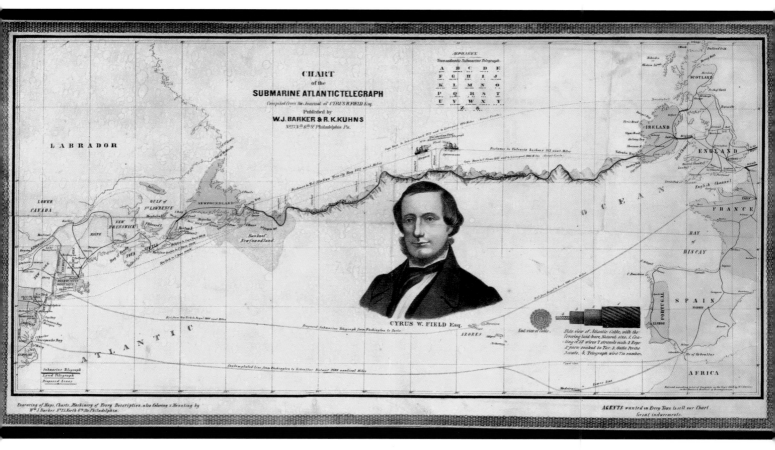

Agamemnon and USS Niagara. The plan was that each ship would carry half the cable and meet mid-Atlantic where the cable would be spliced together. The cable broke on the first day, but was repaired. It next broke again in deep waters and the attempt was abandoned.

In July 1858 the Agamemnon, Valorous, Niagara and Gorgon met mid-ocean for the third attempt. The first two ships started laying cable eastwards and Niagara and Gorgon did so westwards. The cable, made up from seven copper wires coated with three coats of gutta percha (a rigid natural latex), wound with tarred hemp and then sheathed in a helix pattern of 18 strands of iron wire, lasted in service for only a few weeks, but that was enough to show that the idea was achievable. The quality of this cable quickly deteriorated and communication became increasingly slower until it stopped altogether in mid-September.

In 1866 the SS Great Eastern succeeded in laying the first permanent cable across the Atlantic from Valentia Island to Trinity Bay, Newfoundland. This time the cable was

▲ A schematic showing the path of the transatlantic telegraph cable. It also shows proposed new cables from Washington to Paris and Washington to Gibraltar. The image was published in 1858 by W.J. Barker of the Eastern District of Pennsylvania. The first message sent was a congratulations telegraph between the directors of the Atlantic Telegraph Company on either side of the Atlantic and the second from Queen Victoria to President James Buchanan on 16 August congratulating him on the enterprise (her 98-word message took 16 hours to send). Library of Congress

composed of seven twisted strands of pure copper coated with Chatterton's compound (an insulating, adhesive waterproof compound), four layers of gutta percha interspaced with layers of adhesive compound, and covered with preservative-saturated hemp and then encased in 18 stands of high-tensile wire wrapped in a spiral pattern. The whole was then wrapped in manila yarn, also steeped in preservative. It took 250 workers eight months of toiling to make the 48,280km of cable. It weighed almost twice as much as the previous cable and took five months to load onto the Great Eastern.

99

SUEZ CANAL
FERDINAND DE LESSEPS – EGYPT – 1859–69

When the Suez Canal opened world trade immediately became much quicker. Within a few years, the canal had an enormous impact on trade generally. It also made European colonisation of Africa much easier.

▼ *This map of the Suez Canal in 1976 shows just how much of the canal was routed through shallow lakes, as it cut across Africa from the Mediterranean Sea to the Gulf of Suez where it ultimately joined the Red Sea. The demilitarised zone, in green, no longer exists.* Library of Congress

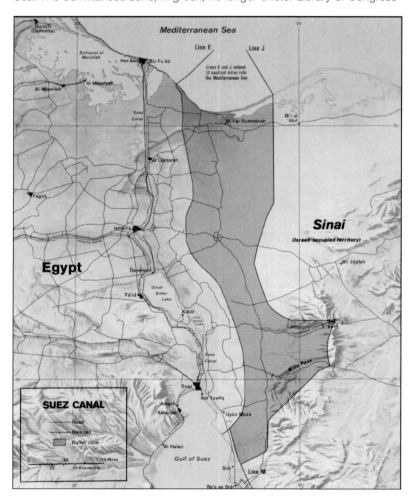

The Suez Canal was built to connect the Mediterranean Sea through the Isthmus of Suez to the Red Sea, a passage of 161km of which about two-thirds is through shallow lakes. The canal was built to reduce the sea-passage time and improve maritime safety between the Atlantic and the Indian oceans, reducing the journey by around 6,920km.

Ancient attempts had been made to dig canals and link the seas and Emperor Napoleon seriously considered doing so for France until put off by the costs of the project. In 1854 Ferdinand de Lesseps (1805–95), a former French consul to Cairo, sealed an agreement with the Ottoman governor of Egypt to build a canal. After extensive surveys and analysis, an international team of engineers drew up a construction plan and the Suez Canal Company was established in December 1858. The cost was estimated to be 200 million francs (on completion this had doubled) and work started with picks and shovels in April the following year at the future Port Said.

Excavating the canal took ten years, mostly through sand, but also some solid rock. It involved up to 30,000 workers from many different countries, many of whom died of cholera and other diseases during the build. Initially the workforce comprised conscripted Egyptians who gave the British government – which had never approved of the project, fearing it would give Britain's European competitors ready access to India – an excuse to interfere and stop the work for a period. But construction resumed, much of it supported by French financial backing and eventually even the British came round to the project.

The first section of canal opened four years behind schedule on 17 November 1869, when the protecting barrage was breached and the waters of the Mediterranean flowed into the Red Sea. There ensued great celebrations, attended by 6,000 people, and an elaborate opening

ceremony led by the Khedive of Egypt and Sudan, accompanied by Empress Eugenie of France in the imperial yacht *L'Aigle*. The second vessel through the canal was the British P&O liner *Delta*.

The initial canal was 22m wide at the bottom and between 60 and 90m wide at the surface but only 7.5m deep.

Financial difficulties meant that the canal was not fully completed until 1871 and even then for the first few years it did not see much traffic. In the first year fewer than 500 ships took the passage, most of them British. In 1876 it was decided to improve the canal by widening and deepening it.

In 1882 the British gained control of the canal. They had by then become the largest shareholders when the British prime minister Benjamin Disraeli bought up the stock of the new Ottoman governor of Egypt for £4 million. In 1888 the Convention of Constantinople declared the canal a neutral zone under the protection of the British, who were occupying Egypt and Sudan at the time.

▲ *Lantern slide view of the Suez Canal taken in about 1914 by William Henry Goodyear (1846–1923). By this time the canal had been widened to take bigger vessels, improving the levels of traffic through it.* Library of Congress

◀ *Passenger steam ship going through the Suez Canal. By the turn of the 19th century pleasure cruises along the Suez Canal became very popular with intrepid travellers.* Library of Congress

100 ROYAL ALBERT BRIDGE
ISAMBARD KINGDOM BRUNEL – DEVON–CORNWALL, ENGLAND – 1859

The Royal Albert Bridge over the River Tamar at Saltash is the crowning – and final – glory of the work of the mercurial Isambard Kingdom Brunel, arguably the most innovative of all Victorian engineers.

Of all the great names of Victorian engineering, one name stands out as the most mercurial: Isambard Kingdom Brunel (1806–59) – if only because it was emblazoned across his last great masterpiece, the Saltash railway bridge across the Tamar between Devon and Cornwall. Born in Portsmouth, the son of émigré French engineer Marc Isambard Brunel (1769–1849) and an Englishwoman, Sophia Kingdom (c1775–1855), Brunel was both a pioneer and a maverick. His choice of a broader gauge – 7ft ¼in – for his railways ran counter to the prevailing use of narrow gauges – most commonly the future standard gauge (4ft 8½in) – and resulted in the necessity of the British government establishing the Royal Commission on Railway Gauges in 1845 to determine what should be the national standard. The subsequent Act of 1846 determined that 4ft 8½in should be the standard gauge in Great Britain and 5ft 3in the standard gauge in Ireland. Although the Great Western Railway – Brunel's mainline railway from London to Bristol and the southwest – was allowed to continue to build new broad-gauge lines, eventually operational practicality dictated that the broad gauge be eliminated and, in 1892, the final sections of 7ft ¼in track were converted.

In extending his railway from Devon into Cornwall, Brunel faced the challenge not only of crossing the Tamar – the 1846 Act authorising construction of the railway stipulated the replacement of the existing ferry by a railway bridge – but also constructing a bridge to satisfy the Admiralty that the Royal Navy's base at Devonport would not be adversely affected.

► *Viewed from the northeast towards Saltash, the scale of Brunel's Royal Albert Bridge is apparent. As with a number of other bridges – such as those across the Menai Strait – the design of the structure was heavily predicated on the demands of the Admiralty that there was sufficient clearance under the bridge to permit Royal Navy vessels to pass safely at high tide.*
Library of Congress

The design of the resulting viaduct – known as the Royal Albert Bridge – was refined on several occasions as a result of Admiralty requirements, the local geology and the railway's lack of funds. The final structure was based around a central pier midstream with two main spans of 139m supported by ten shorter spans on the western side and seven on the eastern. Although Brunel initially thought about suspending the two main spans, he lacked secure accommodation for the tension cables, so opted to use two self-supporting trusses. The original plan had been for the bridge to be built with double track. However, it was decided to accommodate only a single line, thus saving the impecunious railway about £100,000.

In order to construct the central pier in the middle of the Tamar, Brunel drew upon his experience in constructing his father's Thames Tunnel (see p.108). A cylinder with a height of 25.9m and a diameter of 11.25m was constructed. This was floated out to the middle of the river and sunk to act as a coffer dam. With its top sealed, compressed air was then pumped in and, with the water expelled, up to 40 men were able to work at any one time in excavating the mud and bedrock to provide secure foundations. The two bowstring trusses were constructed off site and were floated into position and then gradually raised up to their final height of some 30.5m above sea level by hydraulic jacks. Work on raising the first truss commenced on 1 September 1857 and it was raised to its full height on 1 July 1858. Work on the second truss commenced on 10 July 1858. The completed bridge was first tested on 11 April 1859 and was formally opened on 2 May 1859.

Despite this triumph, work on the bridge, as well as work on his other ventures, took their toll on Brunel's health. He was unable to attend the bridge's opening and he died four months later, on 5 September 1859.

▼ *The Royal Albert Bridge seen from the mouth of the Lynher River. Behind it is the Tamar suspension bridge carrying the A38. Opened in 1962, it was widened and strengthened in 1999–2002.* Author's collection